## *Truth or Truthiness*

Teacher tenure is a problem. Teacher tenure is a solution. Fracking is safe. Fracking causes earthquakes. Our kids are overtested. Our kids are not tested enough.

We read claims like these in the newspaper every day, often with no justification other than "it feels right." How can we figure out what *is* right?

Escaping from the clutches of truthiness begins with one simple question: "what's the evidence?" With his usual verve and flair, Howard Wainer shows how the skeptical mind-set of a data scientist can expose truthiness, nonsense, and outright deception. Using the tools of causal inference he evaluates the evidence, or lack thereof, supporting claims in many fields, with special emphasis in education.

This wise book is a must-read for anyone who's ever wanted to challenge the pronouncements of authority figures and a lucid and captivating narrative that entertains and educates at the same time.

Howard Wainer is a Distinguished Research Scientist at the National Board of Medical Examiners who has published more than four hundred scholarly articles and chapters. This is his twenty-first book. His twentieth book, *Medical Illuminations: Using Evidence, Visualization and Statistical Thinking to Improve Healthcare* was a finalist for the Royal Society Winton Book Prize.

THE DAWN OF REASON

Robert Weber The New Yorker Collection/Cartoon Bank, reproduced with permission.

# Truth or Truthiness

## *Distinguishing Fact from Fiction by Learning to Think Like a Data Scientist*

Howard Wainer

*National Board of Medical Examiners*

**CAMBRIDGE**
**UNIVERSITY PRESS**

32 Avenue of the Americas, New York, NY 10013-2473, USA

Cambridge University Press is part of the University of Cambridge.

It furthers the University's mission by disseminating knowledge in the pursuit
of education, learning, and research at the highest international levels of excellence.

www.cambridge.org
Information on this title: www.cambridge.org/9781107130579



First published 2016

Printed in the United States of America

*A catalog record for this publication is available from the British Library.*

ISBN 978-1-107-13057-9 Hardback

YBP Library Services

WAINER, HOWARD.

TRUTH OR TRUTHINESS: DISTINGUISHING FACT FROM
FICTION BY LEARNING TO THINK LIKE A DATA
SCIENTIST. Cloth 210 P.
NEW YORK: CAMBRIDGE UNIV PRESS, 2016

DISCUSSES USE OF STATISTICAL REASONING & EVIDENCE
IN DECIDING WHICH CLAIMS TO BELIEVE, ETC.
LCCN 2015040736
ISBN 1107130573 Library PO# FIRM ORDERS

8395 NATIONAL UNIVERSITY LIBRAR
App. Date 3/02/16 COLS-PSY 8214-08

List 29.99 USD
Disc 14.0%
Net 25.79 USD

SUBJ: 1. CRITICAL THINKING. 2. INFERENCE. 3.
EVIDENCE.

CLASS BF441 DEWEY# 001.42 LEVEL GEN-AC

*To Linda*
*and*
*Sam & Jennifer*
*and*
*Laurent, Lyn & Koa*

# Annotated Table of Contents

# Preface and Acknowledgments

There have been many remarkable changes in the world over the last century, but few have surprised me as much as the transformation in public attitude toward my chosen profession, statistics – the science of uncertainty. Throughout most of my life the word *boring* was the most common adjective associated with the noun *statistics*. In the statistics courses that I have taught, stretching back almost fifty years, by far the most prevalent reason that students gave for why they were taking the course was "it's required." This dreary reputation nevertheless gave rise to some small pleasures. Whenever I found myself on a plane, happily involved with a book, and my seatmate inquired, "What do you do?" I could reply, "I'm a statistician," and confidently expect the conversation to come to an abrupt end, whereupon I could safely return to my book. This attitude began to change among professional scientists decades ago as the realization grew that statisticians were the scientific generalists of the modern information age. As Princeton's John Tukey, an early convert from mathematics, so memorably put it, "as a statistician, I can play in everyone's backyard."

Statistics, as a discipline, grew out of the murk of applied probability as practiced in gambling dens to wide applicability in demography, agriculture, and the social sciences. But that was only the beginning. The rise of quantum theory made clear that even physics, that most deterministic of sciences, needed to understand uncertainty. The health professions joined in as *Evidence-Based Medicine* became a proper noun. Prediction models combined with exit polls let us go to sleep early with little doubt about election outcomes. Economics and finance was transformed as "quants" joined the investment teams and their success made it clear that you ignore statistical rigor in devising investment schemes at your own peril.

These triumphs, as broad and wide ranging as they were, still did not capture the public attention until Nate Silver showed up and starting predicting the outcomes of sporting events with uncanny accuracy. His success at this gave him an attentive audience for his early predictions of the outcomes of elections. Talking heads and pundits would opine, using their years of experience and deeply held beliefs, but anyone who truly cared about what would happen went to FiveThirtyEight, Silver's website, for the unvarnished truth.

After Nate Silver my life was not the same. The response to my declaration about being a statistician became "Really? That's way cool!" The serenity of long-distance air travel was lost.

As surprising as this shift in attitudes has been, it is still more amazing to me how resistant so many are to accepting evidence as a principal component in deciding between conflicting claims. I chalk this up to three possible reasons:

1. A lack of understanding of the methods and power of the Science of Uncertainty.
2. A conflict between what is true and what is wished to be true.
3. An excessive dimness of mind that prevents connecting the dots of evidence to yield a clear picture of likely outcome.

The first reason is one of my principal motivations in writing this book. The other was my own enthusiasm with this material and how much I want to share its beauty with others.

The second reason was reflected in Upton Sinclair's observation, "It is difficult to get a man to understand something, when his salary depends upon his not understanding it!" We have seen how senators from coal-producing states are against clean air regulations; how the National Rifle Association believes, despite all evidence (see Chapter 11), that more guns will lower the homicide rate; and how purveyors of coastal real estate believe that rising seas accompanying global warming are a pernicious myth.

The third reason is a late addition to my list, and would be unfair, if the observed behavior could be explained with reason two. But I was forced to include it when, on Thursday, February 26, 2015, Senator Jim Inhofe (Republican from Oklahoma, who is Chairman of the Senate Environment and Public Works Committee) brought a snowball onto the floor of the senate as proof that reactions to evidence about global warming are hysterical, and the report that 2014 was the warmest year on record was anomalous. What could explain Senator Inhofe's statements? It could be (1), but as a senator he has been privy to endless discussions by experts with exquisite pedigrees and credentials, which anyone with any wit would be forced to acknowledge as credible. It could be (2) if, for example, his principal supporters were in the petroleum industry, whose future would be grimmer if the nation were to take seriously the role that the burning of such fuels has on global warming. I note that three of the five billionaires in the state of Oklahoma (Harold Hamm, George Kaiser, and Lynn Schusterman) owe their fortunes to the oil and gas industry. That being the case, it isn't surprising that Senator Inhofe might owe some allegiance to the welfare of their financial interests. What makes him a possible candidate for the third category is his apparent belief that his argument would burnish his national reputation rather than making him a punchline on newscasts and late-night TV. I am reminded of Voltaire's prayer "Dear God, make my enemies ridiculous." He knew that politicians can endure anything but the sorts of ridicule that renders them a joke. That Senator Inhofe would purposely place himself into such a position suggests including him in category (3).

Senator Inhofe is not alone on such a list. I would probably want to also include then-senator (now governor) Sam Brownback (Republican from Kansas), former governor Mike Huckabee (Republican from Arkansas), and Representative Tom Tancredo (Republican from Colorado) who, in a 2007 presidential debate all raised their hands to indicate their lack of belief in evolution. It isn't hard to find other possible candidates.[1]

[1] Michelle Bachmann, a six-term congresswoman from Minnesota comes immediately to mind for her avid support of the teaching of creationism in schools, whose appeal to fundamentalist members of her constituency would seem to provide evidence to place her more firmly in (2). Yet, the behavior she exhibited that earned a place on a number of "2013 most corrupt" lists (being under investigation by the Federal Election Commission, House Ethics Committee, and Federal Bureau of Investigation for violating campaign finance laws while running for president by improperly paying staff from her leadership Political Action Committee and using her campaign resources to promote her memoir) suggests she holds a belief in her own invincibility that makes (3) a more likely explanation

I thoroughly understand that no presentation, no matter how lucid, no matter how compelling, can have a direct effect on diminishing either (2) or (3). However, I have hopes that some indirect help can be had by improving the statistical literacy of the general population. The greater the proportion of the population that can detect deceptive, fact-free arguments and hence not be swayed by them, the less effective such arguments will become. I do not believe that this will result in those people whose arguments are based on truthiness changing to another approach. My hopes lie in an educated electorate choosing different people. Paraphrasing Einstein, "old arguments never die, just the people who make them."

I have recently become haunted by lasts. We are usually immediately aware of the firsts in our lives: our first car, first love, first child. We typically become aware of lasts only well after the event; the last time I spoke with my father, the last time I carried my son on my shoulders, the last time I hiked to the top of a mountain. Usually, at least for me, the realization that a last has occurred yields a sense of loss, a deep regret. Had I known it was the last time I would ever speak with my grandfather there are some things I would have liked to have spoken about. Had I known it was the last time I would see my mother I would have told her how much I loved her.

As you read this, I will be well passed the biblically prescribed life span of three-score and ten. Although this is surely my latest book, it might very well be my last. To minimize my future regrets, I want to be especially careful to thank all of those who have contributed both to this specific work and to the more general, and more difficult, task of shaping my thinking.

I begin with my employer, the National Board of Medical Examiners, which has been my professional home since 2001 and has provided a place of peace, quiet, and intellectual stimulation. The modern character of the National Board, a century-old institution, has largely been set by Donald Melnick, its longtime president, whose vision of the organization included room for scholarship and basic research. My gratitude to him and to the organization he leads is immense.

Next, I must thank my colleagues at the National Board beginning with Ron Nungester, senior vice president, and Brian Clauser, vice president, who have always provided support and a thoughtful response to any questions I might have had – both procedural and substantive. In addition, my colleagues Peter and Su Baldwin, Editha Chase, Steve Clyman, Monica Cuddy, Richard Feinberg, Bob Galbraith, Marc Gessaroli, Irina Grabovsky, Polina Harik, Michael Jodoin, Peter Katsufrakis, Ann King, Melissa Margolis, Janet Mee, Arden Ohls, and Peter Scoles have all endured frequent visits in which I either inquired their opinion on some matter of concern to me at the moment, or sat through my explanations of one obscure thing or another. These explanations would typically continue until I decided that I had, at long last, understood what I was talking about. Sometimes this took a long time. I thank them all for their ideas and their forbearance.

Over the course of the past half-century many intellectual debts have accumulated to friends and colleagues who have taught me a great deal. I have neither space nor memory enough to include everyone, but with those limitations in mind, my major benefactors have been: Leona Aiken, Joe Bernstein, Jacques Bertin, Al Biderman, Darrell Bock, Eric Bradlow, Henry Braun, Rob Cook, Neil Dorans, John Durso, Steve Fienberg, Paul Holland, Larry Hubert, Bill Lichten, George Miller, Bob Mislevy, Malcolm Ree, Dan Robinson, Alex Roche, Tom Saka, Sam Savage, Billy Skorupski, Ian Spence, Steve Stigler, Edward Tufte, Xiaohui Wang, Lee Wilkinson, and Mike Zieky.

A very special thinks to David Thissen, my once student, longtime collaborator, and very dear friend.

Next, a mystery. I spent three years as a graduate student at Princeton University acquiring my academic union card. Under ordinary circumstance one would expect that those three years would not have a very different effect on my life than any number of other time periods of similar length. But, that does not seem to have been the case. On a regular basis in the forty-seven years since I left her campus I have been in need of guidance of one sort or another. And unfailingly, before I could flounder for too long, a former Tiger appeared and gave me as much assistance as was needed. Those most prominent in my continued education and productivity are:

John Tukey *39, Fred Mosteller *46, Bert Green *51, Sam Messick *56, Don Rubin ‘65, Jim Ramsay *66, Shelby Haberman ‘67, Bill Berg *67,

Linda Steinberg S*68 P07, Charlie Lewis *70, Michael Friendly *70, Dave Hoaglin *71, Dick DeVeaux '73, Paul Velleman *75, David Donoho '79, Cathy Durso '83, and Sam Palmer '07.

What's the mystery? A quick glance through this list shows that only about four or five were on campus at the same time I was. How did I get to meet the others? For example, I have collaborated with Hoaglin and Velleman for decades and we have yet to be able to recollect where, when, or how we met despite a fair amount of discussion. The only theory that seems to hold water is that Mother Princeton, has decided to take care of her progeny and has somehow arranged to do so. Whatever the mechanism, they and she have my gratitude.

Last, to the staff at Cambridge University Press, the alchemist who has transformed the digital bits and pieces of my manuscript into the handsome volume you hold in your hand now. *Primus inter pares* is my editor, Lauren Cowles, who both saw the value in what I was doing and insisted that I continue rewriting and revising until the result lived up to the initial promise that she had divined. She has my sincerest thanks. In addition, I am grateful for the skills and effort of copy editor Christine Dunn, indexer Lin Maria Riotta and Kanimozhi Ramamurthy and her staff at Newgen Knowledge Works.

# Introduction

*The modern method is to count;*
*The ancient one was to guess.*
Samuel Johnson

In the months leading up to Election Day 2012 we were torn between two very different kinds of outcome predictions. On one side were partisans, usually Republicans, telling us about the imminent defeat of President Obama. They based their prognostication on experience, inside information from "experts," and talking heads from Fox News. On the other side, were "the Quants" represented most visibly by Nate Silver, whose predictions were based on a broad range of polls, historical data, and statistical models. The efficacy of the former method was attested to by snarky experts, armed with anecdotes and feigned fervor, who amplified the deeply held beliefs of their colleagues. The other side relied largely on the stark beauty of unadorned facts. Augmenting their bona fides was a history of success in predicting the outcomes of previous elections, and, perhaps even more convincing, was remarkable prior success, using the same methods, in predicting the outcome of a broad range of sporting events.

It would be easy to say that the apparent supporters of an anecdote-based approach to political prediction didn't really believe their own hype, but were just pretending to go along to boost their own paychecks.[1] And perhaps that cynical conclusion was often true. But how

[1] I am thinking of Upton Sinclair's observation that "it is difficult to get someone to understand something if their paycheck depends on their not understanding it."

are we to interpret the behavior of major donors who continued to pour real money into what was almost surely a rat hole of failure? And what about Mitt Romney, a man of uncommon intelligence, who appeared to believe that in January 2013, he was going to be moving into The White House? Perhaps, deep in his pragmatic and quantitative soul, he knew that the presidency was not his destiny, but I don't think so. I believe that he succumbed to that most natural of human tendencies, the triumph of hope over evidence.

We need not reach into the antics of America's right wing to find examples of humanity's frequent preference for magical thinking over empiricism; it is widespread. Renée Haynes (1906–94), a writer and historian, introduced the useful concept of a *boggle threshold*: "the level at which the mind boggles when faced with some new idea." The renowned Stanford anthropologist Tanya Luhrmann (2014) illustrates the boggle threshold with a number of examples (e.g., "A god who has a human son whom he allows to be killed is natural; a god with eight arms and a lusty sexual appetite is weird."). I would like to borrow the term, but redefine it using her evocative phrase, as the place "where reason ends and faith begins."

The goal of this book is to provide an illustrated toolkit to allow us to identify that line – that place beyond which evidence and reason have been abandoned – so that we can act sensibly in the face of noisy claims that lie beyond the boggle threshold.

The tools that I shall offer are drawn from the field of data science. The character of the support for claims made to the right of the boggle threshold we will call their "truthiness."

> Data science is the study of the generalizable extraction of knowledge from data.
>
> *Peter Naur 1960*

> Truthiness is a quality characterizing a "truth" that a person making an argument or assertion claims to know intuitively "from the gut" or because it "feels right" without regard to evidence, logic, intellectual examination, or facts.
>
> *Stephen Colbert, October 17, 2005*

*Data science* is a relatively recent term coined by Peter Naur but expanded on by statisticians Jeff Wu (in 1997) and Bill Cleveland (in

2001). They characterized data science as an extension of the science of statistics to include multidisciplinary investigations, models and methods for data, computing with data, pedagogy, tool evaluation, and theory. The modern conception is a complex mixture of ideas and methods drawn from many related fields, among them signal processing, mathematics, probability models, machine learning, statistical learning, computer programming, data engineering, pattern recognition and learning, visualization, uncertainty modeling, data warehousing, and high-performance computing. It sounds complicated and so any attempt for even a partial mastery seems exhausting. And, indeed it is, but just as one needn't master solid state physics to successfully operate a TV, so too one can, by understanding some basic principles of data science, be able to think like an expert and so recognize claims that are made without evidence, and by doing so banish them from any place of influence. The core of data science is, in fact, science, and the scientific method with its emphasis on only what is observable and replicable provides its very soul.

This book is meant as a primer on thinking like a data scientist. It is a series of loosely related case studies in which the principles of data science are exemplified. There are only a few such principles illustrated, but it has been my experience that these few can carry you a long way.

*Truthiness*, although a new word, is a very old concept and has long predated science. It is so well inculcated in the human psyche that trying to banish it is surely a task of insuperable difficulty. The best we can hope for is to recognize that the core of truthiness's origins lies in the reptilian portion of our brains so that we can admit its influence yet still try to curb it through the practice of logical thinking.[2]

Escaping from the clutches of truthiness begins with one simple question. When a claim is made the first question that we ought to ask ourselves is "how can anyone know this?" And, if the answer isn't obvious, we must ask the person who made the claim, "what evidence do you have to support it?"

[2] It is beyond my immediate goals to discuss what sorts of evolutionary pressures must have existed to establish and preserve truthiness. For such an in-depth look there is no place better to begin than Nobel Laureate Danny Kahneman's inspired book *Thinking Fast, Thinking Slow*.

Let me offer four examples:

1. Speaking to your fetus in utero is important to the child's development.
2. Having your child repeat kindergarten would be a good idea.
3. Sex with uncircumcised men is a cause of cervical cancer in women.
4. There are about one thousand fish in that pond.

Ideas that lean on truthiness are sometimes referred to as "rapid ideas," for they only make sense if you say them fast.[3] Let us take a slower look at each of these claims in turn.

## Claim 1: Talk to Your Fetus

Let us start with a plan to try to gather the kind of evidence necessary to make such a claim, and then try to imagine how close, in the real world, anyone could get to that ideal study. In order to know the extent of the effect any treatment has on a fetus, we have to compare what happens with that treatment with what would have happened had the fetus not had the treatment. In this situation we must compare the child's development after having regular conversations with its mother with how it would have developed had there been only silence. Obviously the same fetus cannot have both conditions. The problem of assessing the value of an action by comparing its outcome with that of a counterfactual isn't likely to have a solution. Instead we'll have to retreat to making such inferences based on averages within groups, in which we have one group of fetuses subject to the action of interest (being spoken to) and another group in which the alternative was tried (the comparison group). If the two groups were formed through a random process, it becomes plausible to believe that what was observed in the comparison (control) group would have been observed in the treatment group had that group had the control condition.

Next, what is the treatment? How much time is spent conversing with the fetus? What is being discussed? Is it OK to nag? Or instruct? Or is just cooing permissible? And what is the alternative condition? Is

[3] The tendency of stupid ideas to seem smarter when they come at you quickly is known in some quarters as the "Dopeler Effect."

it complete silence? Or just no talk directed solely at the fetus? Does the language matter? What about the correctness of syntax and grammar?

And finally, we need a dependent variable. What is meant by "the child's development"? Is it their final adult height? Or is it the speed with which they acquire language? Their general happiness? What? And how do we measure each child and so be able to make the comparison? And when? Is it at birth? At age one? Five? Twenty?

It seems sensible when confronted with claims like this to ask at least some of these questions. The answers will allow you to classify the claim as based on evidence or truthiness.

I have yet to hear anyone who makes such a claim provide any credible evidence.

## Claim 2: Repeat Kindergarten

The same issues that arise in assessing the evidentiary support for the efficacy of fetal conversations repeat themselves here. How would the child do if not held back? What are the dependent variables that reflect the success of the intervention? Could there ever have been an experiment in which some randomly chosen children were held back and others not? And if this unlikely scenario had actually been followed, how was success judged? If it were on something trivial like height or age in first grade, children held back would be taller and older than those who progressed normally, but that isn't what we care about. We want to know whether the children would be happier if their progress is delayed. Are they reading better in sixth grade than they would have had they not been held back?

It isn't hard to construct a credible theory to support repeating a grade – if a child can't add integers, it makes little sense to move them forward into a class where such skill is assumed, but such decisions are rarely so cut and dried. It is more likely a quantitative decision: "Is this child's skill too low to be able to manage at the next level?" This is knowable, but it requires gathering of evidence. We might display the results of such a study as a graph in which a child's math score in kindergarten is plotted on the horizontal axis and her math score in grade one on the vertical axis. This tells us the relation between performances in the two grades, but it does not tell us about the efficacy of repeating kindergarten. For that we need to know the counterfactual event of what the child's score would have been had she repeated kindergarten. We would need to know how scores compared the first time taking the test with the second time, that is, how she did in first grade after repeating and how she would have done in first grade had she not repeated.

Again, it is possible to construct such an experiment, based on average group performance and random assignment, but the likelihood that any such experiment has ever been performed is small.

Try to imagine the response to your asking about what sort of evidence was used to support a teacher's recommendation that your child should repeat kindergarten. The response would be overflowing with truthiness and rich with phrases like "in my experience" or "I deeply feel."

## Claim 3: Male Circumcision as a Cause of Cervical Cancer

This example was brought to my attention by a student in STAT 112 at the University of Pennsylvania. Each student was asked to find a claim in the popular media and design a study that would produce the necessary evidence to support that claim. Then they were to try to guess what data were actually gathered and judge how close those were to what would be required for a credible conclusion.

The student recognized that the decision to have a baby boy circumcised was likely related to social variables that might have some connection with cervical cancer. To eliminate this possibility, she felt that a sensible experiment that controlled for an unseen connection would need to randomly assign boys to be circumcised or not. She also recognized that women's choice of sex partner might have some unintended connection and so suggested that the matings between men and women should also be done at random. Once such a design was carried out, there would be nothing more to do than to keep track of all of the women in the study for thirty or forty years and count up the frequency of cervical cancer on the basis of the circumcision status of their sex partner. Of course, they would need to keep the same partner for all that time, or we would not have an unambiguous connection to the treatment.

Last, she noted that in the United States about twelve thousand women a year are diagnosed with cervical cancer (out of about 155 million women), or about one case for each thirteen thousand women. So the study would probably need at least half a million women in each of the two experimental groups to allow it to have enough power to detect what is likely to be a modest effect.

Once she had prepared this list of desiderata, she realized that such an experiment was almost certainly never done. Instead, she guessed that someone asked a bunch of women with cervical cancer about the status of their companions and found an overabundance of uncircumcised men. This led her to conclude that the possibilities of alternative explanations were sufficiently numerous and likely to allow her to dismiss the claim.

Is there nothing between a full-randomized experiment and a naïve data collection? In situations where the full experiment is too difficult to perform, there are a number of alternatives, like a case-control study that could provide some of the credibility of a full-randomized experiment, with a vastly more practical format.

Modern science is a complex edifice built on techniques that may not be obvious or even understandable to a layperson. How are we to know that the claims being made are not using credible methods of which we are unaware? I will return to this shortly after illustrating it in the next example.

## Claim 4: Counting Fish in a Pond

"There are about one thousand fish in that pond." How could anyone know that? Did they put a huge net across the pond, capture all the fish, and count them? That sounds unlikely. And so, we may doubt the accuracy of the estimate. But perhaps some scientific methods allow such an estimate. Though it is important to maintain a healthy skepticism it is sensible to ask the person making the claim of one thousand fish what supporting evidence she might have. Had we done so, she might have responded, "We used the method of 'capture-recapture.'" Such jargon requires clarification. And so she expands, "Last week we came here and caught 100 fish, tagged them, and threw them back. We allowed a week to pass so that the tagged fish could mix in with the others and then we returned and caught another 100 fish and found that 10 of them were tagged. The calculation is simple, 10% of the fish we caught were tagged, and we know that in total, 100 were tagged. Therefore there must be about 1,000 fish in the pond."

The use of capture-recapture procedures can be traced back at least to 1783, when the famous French polymath Pierre-Simon Laplace used it to estimate the population of France.[4] This approach is widely used for many purposes; one is to estimate the number of illegal aliens in the United States.

## Coda

The lesson to be learned from these four examples is that skepticism is important, but we must keep an open mind to the possibilities of modern data science. The more we know about it, the better we can design *gedanken* experiments that could yield the evidence that would support the claims made. If we can't imagine one that could work, or if whatever we imagine is unlikely to be practical, we should keep our skepticism, but ask for an explanation, based on science not anecdotes, from the person making the claim. The credibility of the methodology is what tells us how likely the claim is to be on the truthiness side of the boggle threshold.

This book has three parts:

I. **How to think like a data scientist** has, as its centerpiece, a beautiful theory of causality that is used to describe some methods of thinking about claims. In each situation, I illustrate the approach with a real-life claim and its supporting evidence. The questions examined range widely from the causes of happiness; the relation between virtuosos in both music and track; how much has fracking in Oklahoma affected the frequency of earthquakes in that state; and even how to evaluate experimental evidence the collection of which has been censored by death.

II. **How data scientists communicate to themselves and others.** I begin with some theory about the importance of empathy and

[4] Amoros 2014.

effective communication, and then narrow the focus to the communication of quantitative phenomena. The topics include communicating the genetic risks of cancer, the media's use of statistical methods, and the mapping of moral statistics.

III. **The application of these tools of thinking and communicating to the field of education.** Among the topics explored are the surprising trends in student performance over the past few decades, the point of teacher tenure in public schools, and what might have motivated the College Board in 2014 to institute three changes to the SAT.

In each section of this book a series of case studies describe some of the deep ideas of modern data science and how they can be used to help us defeat deception. The world of ideas is often divided into two camps: the practical and the theoretical. Fifty years of experience have convinced me that nothing is so practical as a good theory. The problems associated with making causal inferences lie at the very core of all aspects of our attempts to understand the world we live in, and so there is really no other way to begin than with a discussion of causal inference. This discussion focuses on a very good theory indeed, one that has come to be called "Rubin's Model for Causal Inference" after the Harvard statistician Donald Rubin, who first laid it out forty years ago.

Chapters 1 and 2 provide a brief warm-up, so that, in Chapter 3, we can turn our attention to the rudiments of Rubin's Model and show how it can be used to clarify a vexing chicken-and-egg question. It does this by guiding us to the structure of an experiment, the results of which can settle the issue. In Chapter 4 I expand the applicability of Rubin's Model and show how it casts light into dark corners of scientific inquiry in ways that are surprising. In Chapter 5, we continue on this same tack, by using the fundamental ideas of Rubin's Model to help us design experiments that can answer questions that appear, literally, beyond the reach of empirical solution. After this, the story ebbs and flows, but always with conviction borne of facts. I strive to avoid the passionate intensity that always seems to accompany evidence-starved truthiness.

SECTION I

# Thinking Like a Data Scientist

## Introduction

Not all of data science requires the mastery of deep ideas; sometimes aid in thinking can come from some simple rules of thumb. We start with a couple of warm-up chapters to get us thinking about evidence. In the first, I show how the Rule of 72, long used in finance, can have much wider application. Chapter 2 examines a puzzle posed by a *New York Times* music critic, why are there so many piano virtuosos? By adjoining this puzzle with a parallel one in athletics I unravel both with one simple twist of my statistical wrist. In these two chapters we also meet two important statistical concepts: (1) the value of an approximate answer and (2) that the likelihood of extreme observations increases apace with the size of the sample. This latter idea – that, for example, the tallest person in a group of one hundred is likely not as tall as the tallest in a group of one thousand – although this result can be expressed explicitly with a little mathematics it can be understood intuitively without them and so be used to explain phenomena we encounter every day.

I consider the most important contribution to scientific thinking since David Hume to be Donald B. Rubin's Model for Causal Inference. Rubin's Model is the heart of this section and of this book. Although the fundamental ideas of Rubin's Model are easy to state, the deep contemplation of counterfactual conditionals can give you a headache. Yet the mastery of it changes you. In a very real sense learning this approach to causal inference is closely akin to learning how to swim or how to read.

They are difficult tasks both, but once mastered you are changed forever. After learning to read or to swim, it is hard to imagine what it was like not being able to do so. In the same way, once you absorb Rubin's Model your thinking about the world will change. It will make you powerfully skeptical, pointing the way to truly find things out. In Chapter 3 I illustrate how to use this approach to assess of the causal effect of school performance on happiness as well as the opposite: the causal effect happiness has on school performance. And then in Chapter 4 I expand the discussion to show how Rubin's Model helps us deal with the inevitable situation of unanticipated missing data. The example I use is when subjects in a medical experiment die prior to the completion of the study. The results are surprising and subtle.

Fundamental to the power of Rubin's Model is the control available through experimentation. In Chapter 5 I describe several vexing questions in educational testing, where small but carefully designed and executed experiments can yield unambiguous answers, even though prior observational studies, even those based on "Big Data," have only obscured the vast darkness of the topic.

But sometimes practical constraints get in the way of doing experiments to answer important causal questions. When this occurs we are forced to do an observational study. In Chapter 6 we illustrate one compelling instance of this as we try to estimate the size of the causal effect of some modern methods for oil and gas extraction (hydraulic fracturing – fracking – and the high-pressure injection of waste water into disposal wells) on seismic activity. We show how the credibility of our estimates increases the closer we come in our observational study to the true experiment that was not practical.

And finally, we take on what is one of the most profound problems in all practical investigations: missing data. A discussion of what is generally thought to be such an arcane topic is bound to be greeted with yawns of disinterest. Yet it cannot be ignored. In Chapter 7 I show two situations where the way that missing data were treated has yielded huge errors. In both cases the results were purposely manipulated by taking advantage of the quirks in the missing data algorithm that was chosen. In the first case, some of the manipulators lost their jobs, in the second some went to jail. The point of the chapter is to emphasize how important an awareness of the potential for deception provided by missing data is. Once we understand the mischief that can be perpetrated we are motivated to learn to focus our attention on how the inevitable missing data are treated.

1

# How the Rule of 72 Can Provide Guidance to Advance Your Wealth, Your Career, and Your Gas Mileage

*The sciences do not try to explain, they hardly even try to interpret, they mainly make models. By a model is meant a mathematical construct which, with the addition of certain verbal interpretations, describes observed phenomena. The justification of such a mathematical construct is solely and precisely that it is expected to work.*

John Von Neumann

Great news! You have won a lottery and you can choose between one of two prizes. You can opt for either:

1. $10,000 every day for a month, or
2. One penny on the first day of the month, two on the second, four on the third, and continued doubling every day thereafter for the entire month.

Which option would you prefer?

Some back-of-the envelope calculations show that after ten days option (1) has already yielded $100,000, whereas option (2) only yielded $10.23. The choice seems clear, but we continue with some more arithmetic and after twenty days (1) has ballooned up to $200,000, and option (2) has yielded $10,485.75. Is there any way that over the remainder of the month the tortoise of option (2) can possibly overtake the hare of option (1)?

Quietly, however, even after twenty days the exponential momentum has become inexorable, for by day twenty-one it is $21,971, by day twenty-two it is $41,943, and so by day twenty-five, even though option (1) has reached its laudable, but linear, total of $250,000, option (2) has passed it, reaching $335,544 and is sprinting away toward the end-of-the-month finish line.

If the month was a non–leap year February, option (2) would yield $2,684,354, almost ten times option (1)'s total. But with the single extra day of a leap year it would double to $5,368,709. And, if you were fortunate enough to have the month chosen being one lasting thirty-one days, the penny-a-day doubling would have accumulated to $21,474,836.47; almost seventy times the penurious $10,000/day's total.

As we can now see, the decision of which option to choose is not even close. Yet, even though the choice of option (2) was a slam dunk, how many of us could have foreseen it?

Exponential growth has befuddled human intuition for centuries. One of the earliest descriptions of the confusion it engendered was described in the *Shahnameh*, an epic poem by the Persian poet Ferdowski at around 1000 CE. The story revolves around the Indian mathematician Sessa, who invented the game of chess and showed it to his Maharajah. The Maharajah was so delighted with the new game that he gave Sessa the right to name his own prize for the invention. Sessa asked the Maharajah for his prize to be paid in wheat. The amount of wheat was to be determined by placing one grain on the first square of the chessboard, two on the second square, four on the third, and so forth, doubling each time for all sixty-four squares. This seemed to the Maharajah to be a modest request, and he quickly agreed and ordered his treasurer to calculate the total amount and hand it over to Sessa.[1] It turned out that the Maharajah had a far poorer understanding of exponential growth than did Sessa, for the total was so great that it would take more than all the assets of the kingdom to satisfy. There are two versions of the end of the story. In one, Sessa becomes the new ruler; in another, Sessa was beheaded.

[1] The exponential growth, invisible to the Maharajah, would yield 18,446,744,073,709,551,615 grains of wheat.

But we don't have to reach back to preclassical antiquity to find examples of such confusion. In 1869 the British polymath Francis Galton (1822–1911) was studying the distribution of height in Britain. Using the power of the normal distribution Galton was able to project the distribution of heights for the entire population[2] from his modest sample. Because he misestimated how quickly the exponential nature of the normal curve forces it to fall toward zero, he predicted that several occupants of the British Isles would be taller than nine feet. In Galton's time a nine-footer would have been about thirteen standard deviations above the mean, an event of truly tiny likelihood. Without doing the calculations you can check your own intuition by estimating the height of the normal curve in the middle if the height thirteen standard deviations away is but one millimeter. If your answer is measured in any unit smaller than the light year, your intuition is likely as flawed as Galton's.[3]

Compound interest yields exponential growth, so financial planners emphasize the importance of starting to save for retirement as early as possible. And yet the result of the exponential growth yielded by compound interest is hard to grasp, deep in one's soul. To aid our intuition a variety of rules of thumb have been developed. One of the best known, and the oldest, is the "Rule of 72," described in detail (although without derivation) by Luca Pacioli (1445–1514) in 1494.

In brief, the Rule of 72 gives a good approximation as to how long it will take for your money to double at any given compound interest rate. The doubling time is derived by dividing the interest rate into seventy-two. So at 6 percent your money will double in twelve years, at 9 percent in eight years, and so forth.[4] Although this approximation is easy to compute in your head, it is surprisingly accurate (see Figure 1.1).

But exponential growth happens in many circumstances outside of finance. When I was a graduate student, the remarkable John Tukey

[2] In Chapter 2 we will use a corrected version of Galton's method to answer an allied vexing question.

[3] The height in the center of this normal curve is approximately 3.4 million times the diameter of the universe.

[4] The Rule of 72 is a rough approximation. A more accurate one for a doubling in T time periods at an interest rate r is $T = \ln(2)/r$; for tripling it is $T = \ln(3)/r$, etc. Note that $100 \times \mathrm{Ln}(2) = 69.3$, which would provide a slightly more accurate estimate than 72; but because 72 has so many integer factors, it is easy to see why it has been preferred.

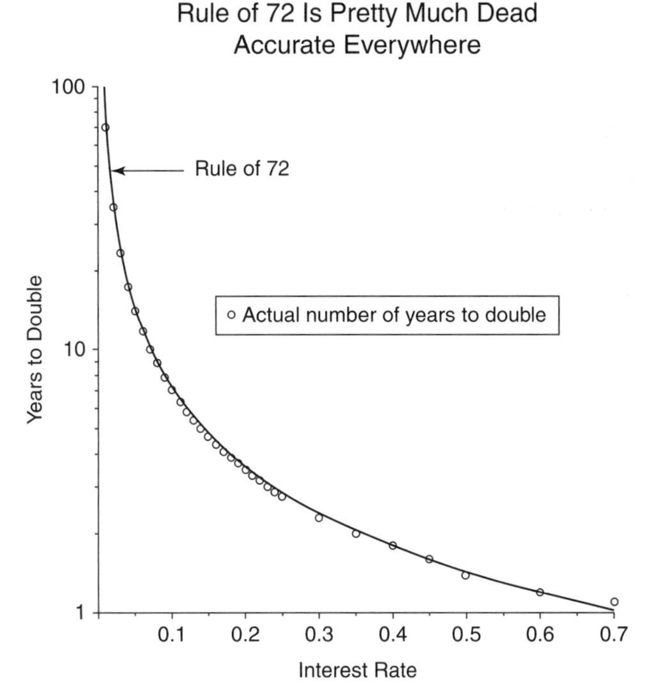


FIGURE 1.1. The power of compound interest shown as the number of years for money to double as a function of the interest rate.

advised me that to succeed in my career, I would have to work harder than my competitors, but "not a lot harder, for if you work just 10% harder in just 7 years you will know twice as much as they do." Stated in that way, it seems that at a cost of just forty-eight minutes a day you can garner huge dividends.

Now that we have widened our view of the breadth of application of the Rule of 72, we can easily see other avenues where it can provide clarity. For example, I recently attended a fiftieth high school reunion and was dismayed at the state of my fellow graduates. But, once I realized that those who had allowed their weight to creep upward at even the modest rate of 1.44 percent a year would, at the fiftieth reunion, be double the size I remembered from their yearbook portrait.

In the same way, if we can increase gas mileage just 4 percent each year, in only eighteen years the typical new car would use only half as much gasoline as today's cars.

Of course, this rule also provides insight into how effective various kinds of plans for world domination can be affected. One possible way for a culture to dominate all others is for its population to grow faster than its competitors. But not a great deal faster; for again, if the growth rate is just 6 percent greater its population will double in just twelve years.

Here I join with Mark Twain (1883) in that what we both like best about science is that "one gets such wholesale returns of conjecture out of such a trifling investment of fact."

2

# Piano Virtuosos and the Four-Minute Mile

"Virtuosos becoming a dime a dozen," exclaimed Anthony Tommasini, chief music critic of the *New York Times* in his column in the arts section of that newspaper on Sunday, August 14, 2011. Tommasini described, with some awe, the remarkable increase in the number of young musicians whose technical proficiency on the piano allows them to play anything. He contrasts this with some virtuosos of the past – he singles out Rudolf Serkin as an example – who had only the technique they needed to play the music that was meaningful to them. Serkin did not perform pieces like "Prokofiev's finger twisting Third Piano Concerto or the mighty Liszt Sonata," although such pieces are well within the capacity of most modern virtuosos.[1]

But why? Why have so many young pianists set "a new level of conquering the piano"? Tommasini doesn't attempt to answer this question (although he does mention Roger Bannister in passing), so let me try.

[1] A new level of technical excellence is expected of emerging pianists. I see it not just on the concert circuit but also at conservatories and colleges. In recent years I have repeatedly been struck by the sheer level of instrumental expertise that seems a given.

The pianist Jerome Lowenthal, a longtime faculty member at Juilliard School of Music, observes it in his own studio. When the 1996 movie *Shine*, about the mentally ill pianist David Helfgott, raised curiosity about Rachmaninoff's Third Piano Concerto, Mr. Lowenthal was asked by reporters whether this piece was as formidably difficult as the movie had suggested. He said that he had two answers: "One was that this piece truly is terribly hard. Two was that all my 16-year-old students were playing it." *Anthoni Tommasini,* New York Times, *August 14, 2011.*

We see an apparently unending upward spiral in remarkable levels of athletic achievement that provides a context to consider Tommasini's implicit riddle. I don't mean to imply that this increase in musical virtuosity is due to improved diet and exercise, or even to better coaching; although I would be the last to gainsay their possible contribution. I think a major contributor to this remarkable increase in proficiency is population size. I'll elaborate.

The world record for running the mile has steadily improved by almost four-tenths of a second a year for the past century. When the twentieth century began the record was 4:13. It took almost fifty years until Roger Bannister collapsed in exhaustion after completing a mile in just less than four minutes. In a little more than a decade his record was being surpassed by high school runners. And, by the end of the twentieth century, Hicham El Guerrouj broke the tape at 3:43. What happened? How could the capacity of humans to run improve so drastically in such a relatively short time? Humans have been running for a very long time, and in the more distant past, the ability to run quickly was far more important for survival than it is today. A clue toward an answer lies in the names of the record holders. In the early part of the century the record was held by Scandinavians – Paavo Nurmi, Gunder Haag, and Arne Andersson. Then mid-century came the Brits: Roger Bannister, John Landy, Herb Elliot, Peter Snell, and later Steve Ovett and Sebastian Coe. And in the twenty-first century the Africans arrived; first Filbert Bayi, then Noureddine Morceli and Hicham el Guerrouj. As elite competition began to include a wider range of runners, times improved. A runner who wins a race that is the culmination of events that winnowed the competition from a thousand to a single person is likely to be slower than one who is the best of a million.

A simple statistical model, proposed and tested in 2002 by Scott Berry, captures this idea. It posits that human running ability has not changed over the past century. That in both 1900 and 2000 the distribution of running ability of the human race is well characterized by a normal curve with the same average and the same variability. What has changed is how many people live under that curve. And so in 1900 the best miler in the world (as far as we know) was the best of a billion; in 2000 he was the best of six billion. It turns out that this simple model can accurately describe the improvements in performance of all athletic contests for which there is an objective criterion.[2]

It does not seem far-fetched to believe the same phenomenon is taking place in other areas of human endeavor. Looking over the list of extraordinary young pianists mentioned by Tommasini, we see names that are commonplace now, but would have seemed wildly out of place at Carnegie Hall a century ago – Lang Lang, Yundi Li, and Yuja Wang. As the reach of classical music extended into areas previously untouched, is it any surprise that among the billions of souls who call such places home we would discover some pianists of remarkable virtuosity?

Tommasini illustrates his point with his reaction to eighty-year-old recordings of the respected pianist Alfred Cortot. He concludes that Cortot "would probably not be admitted to Julliard now." This should not surprise us any more than the likelihood that Paavo Nurmi, the flying Finn, would have trouble making a Division I collegiate track team. The best of a billion is almost surely better than the best of a million.

[2] Of course, social factors – the shrinking and homogenization of the world – have increased the effect still more. Kenya did not compete in the 1900 games – nor did any other African country. Not only was the global pool smaller, but an entire continent's population was excluded from it. At that time probably none of the long-legged Kalenjin peoples who lived and still live at high altitude seven thousand feet up on the edge of the Rift Valley had ever heard of the Olympics, or even of Paris, where they were held that year; and if they had, the practical as well as the conceptual possibility of traveling across a quarter of the world just to run a race simply did not exist. Kenya did not compete in the Olympics until 1956; three games later their athletes picked up three gold medals in Mexico City – the famous Kip Keino was only one of those athletes. Since then Kenyans, the vast majority of them Kalenjin, have gathered sixty-eight Olympic medals in athletics. Noah Beardsley has calculated that in 2005 the Kalenjin made up .0005 percent of the world's population but won 40 percent of the top international distance running events.

# 3

# Happiness and Causal Inference[1]

## Introduction

My old, and very dear, friend Henry Braun describes a data scientist as someone who's pretty good with numbers but hasn't got the personality to be an accountant. I like the ambiguity of the description, vaguely reminiscent of a sign next to a new housing development near me, "Never so much for so little." But although ambiguity has an honored place in humor, it is less suitable within science. I believe that although some ambiguity is irreducible, some could be avoided if we could just teach others to think more like data scientists. Let me provide one illustration.

Issues of causality have haunted human thinkers for centuries, with the modern view usually ascribed to the Scot David Hume. Statisticians Ronald Fisher and Jerzy Neyman began to offer new insights into the topic in the 1920s, but the last forty years, beginning with Don Rubin's unlikely sourced 1974 paper, have witnessed an explosion in clarity and explicitness on the connections between science and causal inference. A signal event in statisticians' modern exploration of this ancient topic was Paul Holland's comprehensive 1986 paper "Statistics and Causal Inference," which laid out the foundations of what he referred to as "Rubin's Model for Causal Inference."

[1] My gratitude to Don Rubin for encouragement and many helpful comments and clarifications.

Causa latet: vis est notissima[2]

*Ovid, Metamorphosis, IV c. 5*

A key idea in Rubin's model is that finding the cause of an effect is a task of insuperable difficulty, and so science can make itself most valuable by measuring the effects of causes. What is the effect of a cause? It is the difference between what happens if some unit is exposed to some treatment versus what would have been the result had it not been.[3] This latter condition is a counterfactual and hence impossible to observe. Stated in a more general way, the causal effect is the difference between the actual outcome and some unobserved potential outcome.

Counterfactuals can never be observed hence, for an individual, we can never calculate the size of a causal effect directly. What we can do is calculate the average causal effect for a group. This can credibly be done through randomization. If we divide a group randomly into a treatment group and a control group (to pick one obvious situation), it is credible to believe that, because there is nothing special about being in the control group, the result that we observe in the control group is what we would have observed had the treatment group been enrolled in the control condition. Thus the difference between the treatment and the control outcomes is a measure of the size of the average causal effect of the treatment (relative to the control condition). The randomization is the key to making this a credible conclusion. But, in order for randomization to be possible, we must be able to assign either treatment or control to any particular participant. Thus is derived Rubin's[4] bumper-sticker-worthy conclusion that there can be "no causation without manipulation."

This simple result has important consequences. It means that some variables, like gender or race, cannot be fruitfully thought of as causal because we cannot randomly assign them. Thus the statement "she is short because she is a woman" is causally meaningless, for to measure the

2 The cause is hidden, but the effect is known.

3 Hume famously defined causality: "We may define a cause to be an object followed by another, and where all the objects, similar to the first, are followed by objects similar to the second, and, where, if the first object had not been, the second would never have existed." It is the underlined second clause where counterfactuals enter the discussion. (1740, Author emphasis.)

4 Rubin 1975, 238.

FIGURE 3.1. Dilbert cartoon.
*Source*: Courtesy of AMU Reprints.

effect of being a woman we would have to know how tall she would have been had she been a man. The heroic assumptions required for such a conclusion removes it from the realm of empirical discussion.

Although the spread of Rubin's and Holland's ideas has been broad within the statistical community, their diffusion through much of the social sciences and the humanities, where they are especially relevant, has been disappointingly slow. The one exception to this is in economics, where making valid causal inferences has crucial importance. One goal of this chapter is to help speed that diffusion by showing how they can illuminate a vexing issue in science, assessing the direction of the causal arrow. Or, more precisely, how we can measure the size of the causal effect in each direction. This issue arose, most recently, in an article[5] in the *Journal of the American Medical Association* that proposed a theory of obesity that turned the dominant theory on its head. Specifically, the authors argued that there is evidence that people eat too much because they are fat; not that they are fat because they eat too much. Obviously, measuring the relative size of the effects of the two plausible causes is of enormous practical consequence. Today let us tackle a different manifestation of the same problem, which has some subtler aspects worth illuminating – the effect of performance on happiness and the effect of happiness on performance (see Figure 3.1).

## Happiness: Its Causes and Consequences

There is an extensive literature surrounding the relationship between a human sense of well-being (what I will call "happiness") and successful

[5] Ludwig and Friedman, May 16, 2014.

performance on some cognitive task (say school grades or exam scores). Some observers have suggested that happy students do better in school (e.g., the effect of being happy is higher grades); others point out that when someone does well, it pleases them and they are happier (e.g., the effect of doing well is increased happiness). How are we to disentangle this chicken and egg problem? Before we tackle this, let me drop back a little and describe the state of the art (as much as I could discern it) in the happiness literature.

Some claim[6] that often the rigors associated with high performance generates unhappiness. To achieve the goal of making our children happier, this "finding" has led to the suggestion that academic standards should be relaxed. The existence of this suggestion,[7] and that it is being seriously considered, lifts the subject of the direction of the causal arrow (as well as the size of the causal effect) out of the realm of the theoretical into that of the practically important.

Empirical evidence actually shows a positive relationship between happiness and performance. How credible is this evidence? This is hard for me to judge because much of it appears in scientific journals like the *Journal of Happiness Studies* or *Education Research International* whose scientific rigor is unknown to me. I did note a fair number of cross-sectional studies that show a positive relationship between happiness and school success.[8] But they often carry a caveat akin to:

> As with any study based on correlational evidence, care must be taken in the interpretation and generalization of the findings. Specifically, the nature of the evidence does not support a causal link between examined variables even if one, in truth, does exist. Additional research would therefore be warranted to further investigate additional relational dimensions between and among the variables explored in the present study.[9]

What sort of care? Happily, the authors help us with an elaboration:

> a larger sample with a more even distribution of gender and race could also stand to strengthen the findings as would a

6 Robbins 2006.

7 This notion falls into the category of a "rapid idea." Rapid ideas are those that only make sense if you say them fast.

8 Gilman and Huebner 2006; Verkuyten and Thijs 2002.

9 Walker et al. 2008.

sample of participants from beyond the Midwestern United States and from larger universities.

Is the character of the sample the only problem? In 2007 Quinn and Duckworth point out that the causal questions of interest would be better explored "[i]n a prospective, longitudinal study," which they did. The value of a longitudinal study harks back to Hume's famous criteria for causality. A key one is that a cause must come before an effect. Without gathering longitudinal data we cannot know the order. But this is a necessary condition for causality, not a sufficient one. In Quinn and Duckworth's study they measured the happiness of a sample of students (along with some background variables) and then recorded the students' school grades; they then returned a year later and did it again. They concluded, "Participants reporting higher well-being were more likely to earn higher final grades" and "students earning higher grades tended to go on to experience higher well- being." They sum up: "The findings suggest the relationship between well-being and academic performance may be reciprocally causal."

Trying to draw longitudinal inferences from cross-sectional data is a task of great difficulty. For example, I once constructed a theory of language development from a detailed set of observations I made while on a walking tour of south Florida. I noted that most people, when they were young, primarily spoke Spanish. But, when people were old they usually spoke Yiddish. I tested this theory by noting that the adolescents who worked in local stores spoke mostly Spanish, but a little Yiddish. You could see the linguistic shift happening.

It is easy to see that the results obtained from a longitudinal study are less likely to suffer from the same artifacts as a cross-sectional one. But because a longitudinal study's causal conclusions suffer from fewer possible fatal flaws than a cross-sectional study does not mean that such conclusions are credible. Something more is needed. We turn to Rubin's Model for help.

Let us start with the idea that when someone does well they are happier than if they did badly. Simultaneously, it does not seem far-fetched to believe that happier people will do better. This correlational result does not conflict with Quinn and Duckworth's causal conclusion. But the important question is quantitative not qualitative. Can we design an experiment in which the treatments (e.g., happiness) can be randomly assigned?

Suppose we take a sample of students and randomly divide them in half, say into groups A and B. We now measure their happiness using whatever instruments are generally favored. Next we administer an exam to them and subtract fifteen points from the scores that we report to all students in group A while adding fifteen points to the scores reported to those in group B (it is easy to see generalizations of this in which the size of the treatments are modified in some systematic way, but that isn't important right now). Now we remeasure their happiness. My suspicion is that those students who did worse than they expected would be unhappier than they were originally, and those who did much better will be happier. The amount of change in happiness is the causal effect of the treatment, relative to the control. Had a fuller experiment been done we could easily calculate the functional relationship between the number of points added and the change in happiness. This ends Stage 1 of the experiment.

Next Stage 2: we now have two groups of students whose happiness was randomly assigned, so we can now readminister a parallel form of the same test. We then calculate the difference in the scores from the first administration (the actual score, not the modified one) to the second. The size of that difference is a measure of the causal effect of happiness on academic performance. The Stage 2 portion of the experiment, where the assignment probabilities depend on the outcomes of the first stage, is usually called a sequentially randomized or "split plot" design.[10]

The ratio of the size of the two causal effects tells us the relative influence of the two treatments. An outcome from such an experiment could yield conclusions like "the effect of improved performance has ten times the effect on happiness, than a similar increase in happiness has on performance."[11] Such a quantitative outcome is surely more satisfying

[10] Rubin 2005.

[11] Intuitively this surely makes sense, for if you don't know the answer to a question, being happier isn't going to change your knowledge base.

and more useful than continued speculation about the direction of the causal arrow.

## Conclusions

Even a cursory reading of the happiness literature reveals the kind of conclusions that researchers would like to make. Typically, Zahra, Khak, and Alam (2013, 225) tell us that "[r]esults showed that in addition to the positive and significant correlation between happiness and academic achievement of university students, happiness could also explain 13% of changes of academic achievement." You can feel the authors' desire to be causal, and they come very close to making a causal claim – it's certainly possible to interpret *explain* that way. But the character of the studies done militates against the sorts of causal interpretations that seem to be yearned for. Most were observational studies, and the rest might be called "some data I found lying in the street." But I bring good news. Through the use of Rubin's Model we can design true experimental studies that can provide answers to the questions we want to ask. Moreover, the very act of precisely describing what the real or hypothetical randomized experiment that would be needed to measure the causal effects of interest greatly clarifies what causal question is being asked.

The bad news is that such studies are not as easy as picking up old data off the street and summarizing them; but if making causal inferences correctly were easy everyone would do it.[12]

[12] Precision is important, note that the treatment in Stage 1 is not higher performance versus lower performance, but rather higher scores than expected versus lower scores than expected. A subtle, but important, distinction.

# 4

# Causal Inference and Death

*The best-laid schemes o' mice an' men*
*Gang aft agley,*
*An' lea'e us nought but grief an' pain,*
*For promis'd joy!*
Robert Burns 1785

In Chapter 3 we learned how being guided by Rubin's Model for Causal Inference helps us design experiments to measure the effects of possible causes.[1] I illustrated this with a hypothetical experiment on how to unravel a causal puzzle of happiness. Is it really this easy? The short answer is, unfortunately, no. But in the practical world, more complicated than the one evoked in my proposed happiness study, Rubin's Model is even more useful. In this chapter we go deeper into the dimly lit practical world, where participants in our causal experiment drop out for reasons outside our control. I show how statistical thinking in general, and Rubin's Model in particular, can illuminate it. But let us go slowly and allow time for our eyes to acclimate to the darkness.

Controlled experimental studies are typically regarded as the gold standard for which all investigators should strive, and observational studies as their polar opposite, pejoratively described as "some data we found lying on the street." In practice they are closer to each other than we

[1] This chapter is a lightly rewritten version of Wainer and Rubin (2015). It has benefited from the help and comments from Peter Baldwin, Stephen Clyman, Peter Katsufrakis, Linda Steinberg, and *primus inter pares*, Don Rubin, who generously suggested the area of application and provided the example.

are often willing to admit. The distinguished statistician Paul Holland, expanding on Robert Burns, observed that

> All experimental studies are observational studies waiting to happen.[2]

This is an important and useful warning to all who are wise enough to heed it. Let us begin with a more careful description of both kinds of studies:

The key to an experimental study is control. In an experiment, those running it control:

(1) What is the treatment condition,
(2) What is the alternative condition,
(3) Who gets the treatment,
(4) Who gets the alternative, and
(5) What are the outcome (dependent) variables.

In an observational study the experimenter's control is not as complete. Consider an experiment to measure the causal effect of smoking on life expectancy. Were we to do an experiment, the treatment might be a pack of cigarettes a day for one's entire life. The alternative condition might be no smoking. Then we would randomly assign people to smoke or not smoke, and the dependent variable would be their age at death.

Obviously, such an experiment is theoretically possible, but not practical because we cannot assign people to smoke or not at random.[3] This is the most typical shortcoming of even a well-designed observational study. Investigators could search out and recruit treatment participants who smoke a pack a day, and also control participants who are nonsmokers. And they could balance the two groups on the basis of various observable characteristics (e.g., have the same sex/racial/ethnic/age mix in both groups). But when there is another variable related to length of life that is

[2] Holland, P. W., Personal communication, October 26, 1993.

[3] This experiment has been attempted with animals, where random assignment is practical, but it has never shown any causal effect. This presumably is because easily available experimental animals (e.g., dogs or rats) do not live long enough for the carcinogenic effects of smoking to appear, and animals with long-enough lives (e.g., tortoises) cannot be induced to smoke.

not measured, there may be no balance on it. Randomization, of course, provides balance, on average, for all such "lurking missing variables."

This example characterizes the shortcoming of most observational studies – the possibility of lack of balance between the alternative and treatment conditions that randomization achieves (on average) in an experiment. It also makes clear why an observational study needs to collect lots of ancillary information about each participant so that the kind of balancing required can be attempted. In a true experiment, with random assignment, such information is (in theory) not required. Here enters Paul Holland, whose observations about the inevitability of missing data will further illuminate our journey.

Suppose some participants drop out of the study. They might move away, stop showing up for annual visits, decide to quit smoking (for the treatment condition) or start (for the control), get hit by a truck, or expire for some other reason unrelated to smoking. At this point the randomization is knocked into a cocked hat and we must try to rescue the study using the tools of observational studies. Such a rescue is impossible if we have not had the foresight to record all of the crucial ancillary information, the covariates, that are the hallmark of a good observational study.

Our inability to randomize has allowed missing variables to mislead us in the past. For a long time the effect of obesity on life expectancy was underestimated because smokers tended to both die younger and be more slender than nonsmokers. Hence the negative effect of smoking on life expectancy was conflated with the advantage of avoiding obesity. Experiments in which we randomly assign people to be obese or not are not possible. Modern research on the effects of obesity excludes participants who have ever smoked.

> I used to be Snow White, but I drifted.
>
> *Mae West*

It is obvious now, if it wasn't already, that even the most carefully planned experiments with human beings get contaminated when carried out in the real world. And yet our goals for these experiments do not change. The issue is clear: How can we estimate causal effects when unexpected events intrude on the data collection, causing some observations to be missing?

Let us begin with a fundamental fact[4] that no amount of verbal or mathematical legerdemain can alter.

## The Magic of Statistics Cannot Put Actual Numbers Where There Are None

And so, when we run an experiment and some of the subjects drop out, taking with them the observations that were to be the grist for our inferential mill, what are we to do?

There are many ways to answer this question, but all that are credible must include an increase in the uncertainty of our answer over what it would have been had the observations not disappeared.

## Coronary Bypass Surgery: An Illuminating Example

What can be done when, because of clogged arteries that have compromised its blood supply, the heart no longer operates effectively? For more than fifty years one solution to this problem has been coronary bypass surgery. In this procedure another blood vessel is harvested and sewn into the heart's blood supply, bypassing the clogged one. This surgery carries substantial risks, usually greater for a patient who is not in robust health. Before recommending this procedure widely, it is crucially important to assess how much it is likely to help each type of patient. So let us construct an experiment that would measure the size of the causal effects that such surgery has on the health of the patients after the treatment.

(1) **Treatment** – Bypass surgery in addition to medication, diet, and a program of exercise.
(2) **Control** – Medication, diet, and a program of exercise, but *no* surgery.
(3) **Subjects** in both groups chosen at random from a pool of individuals judged to have one or more clogged arteries that supply the heart.

[4] I suspect that this has been stated many times by many seers, but my source of this version was that prince of statistical aphorisms, Paul Holland (October 26, 1993).

(4) **Outcome Measure** – We shall judge the success of the intervention by a "quality of life" (QOL) score based on both medical and behavioral measures taken one year after the surgery.

After the experiment has been set up, we look carefully at the effectiveness of the randomization to assure ourselves that the two groups match on age, sex, SES (Socio-Economic Status), smoking, initial QOL, and everything else we could think of. It all checks out, and the experiment begins.

> Der mentsh trakht und Got lakht.[5]
>
> *Ancient Yiddish proverb*

As the experiment progresses, some of the patients die. Some died before the surgery, some during, and some afterward. For none of these patients was the one-year QOL measured. What are we to do?

One option might be to exclude all patients with missing QOL from the analysis and proceed as if they were never in the experiment. This approach is too often followed in the analysis of survey data in which the response rate might be just 20 percent, but the results are interpreted as if they represented everyone. This is almost surely a mistake, and the size of the bias introduced is generally proportional to the amount of missingness. One survey by a company of its employees tried to assess "engagement with the company." They reported that 86 percent of those responding were "engaged" or "highly engaged." They also found that only 22 percent of those polled were engaged enough to respond, which at minimum should cause the reader to wonder about the degree of engagement of the 78 percent who opted not to respond.

A second option might be to impute a value for the missing information. For the nonrespondents in the engagement survey one might impute a value of zero engagement for the nonrespondents, but that might be too extreme.

It is tempting to follow this approach with the bypass surgery and impute a zero QOL for everyone who is dead. But is this true? For many people being alive, regardless of condition, is better than the alternative. Yet, a fair number would disagree and would score "being dead"

[5] Man plans and God laughs.

TABLE 4.1. *The Observed Data*

| Percent of Sample | Experimental Assignment | Outcome | |
|---|---|---|---|
| | | Intermediate | Final(QOL) |
| 30 | Treatment | Lived | 700 |
| 20 | Treatment | Died | * |
| 20 | Control | Lived | 750 |
| 30 | Control | Died | * |

higher than some miserable living situations and construct living wills to enshrine this belief.

Let us begin our discussion with an analysis of artificial (but plausible) data from such an experiment,[6] which we summarize in Table 4.1.

From this table we can deduce that 60 percent of those patients who received the treatment lived, whereas only 40 percent of those who received the control survived. Thus, we would conclude that on the intermediate, but still very important, variable of survival, the treatment is superior to the control. But, among those who lived, QOL was higher for those who received the control condition than for those who received the treatment (750 vs. 700) – perhaps suggesting that treatment is not without its costs.

We could easily have obtained these two inferences from any one of dozens of statistical software packages, but this path is fraught with danger. Remembering Picasso's observation that for some things, "computers are worthless; they only give answers." It seems wise to think before we jump to calculate and then to conclude anything from these calculations.

We can believe the life versus death conclusion because the two groups, treatment versus control, were assigned at random. Thus, we can credibly estimate the size of the causal effect of the treatment relative to the control on survival as 60 percent versus 40 percent.

But the randomization is no longer in full effect when we consider QOL, because the two groups being compared were not composed at random, for death has intervened in a decidedly nonrandom fashion and we would err if we concluded that treatment reduced QOL by 50 points. What are we to do?

[6] This example is taken, with only minor modifications, from Rubin 2006.

Rubin's Model provides clarity. Remember one of the key pieces of Rubin's Model is the idea of a potential outcome: each subject of the experiment, before the experiment begins, has a potential outcome under the treatment and another under the control. The difference between these two outcomes would be the causal effect of the treatment, but we only get to observe one of those outcomes. This is why we can only speak of a summary causal effect, for example, averaged over each of the two experimental groups. And this is most credible with random assignment of subjects to treatment.

The bypass experiment has one outcome measure that we have planned, the QOL score, and a second, an intermediate outcome, life or death, which was unplanned. Thus, for each participant we observe whether or not they lived. This means that the participants fall into four conceptual categories (or strata), those who would:

1. Live with treatment and live with control,
2. Live with treatment but die with control,
3. Die with treatment but live with control, or
4. Die with treatment and die with control.

Of course, each of these four survival strata is randomly divided into two groups: those who actually got the treatment and those who got the control.

So for stratum 1 we can observe their QOL score regardless of which experimental group they are in.

In stratum 2, we can only observe QOL for those who got the treatment, and similarly in stratum 3 we can only observe QOL for those who were in the control group. In stratum 4, which is probably made up of very fragile individuals, we cannot observe QOL for anyone.

The summary in Table 4.2 makes it clear that if we simply ignore those who died, we are comparing the QOL of those in cells (a) and (b) with that of those in (e) and (g). We are ignoring all subjects in cells(c), (d), (f), and (h).

At this point three things are clear:

1. The use of Rubin's idea of potential outcomes has helped us to think rigorously about the character of the interpretation of the experiment's results.

TABLE 4.2. *The Experiment Stratified by the Potential Results on the Intermediate (Life/Death) Outcome*

| | Received | |
|---|---|---|
| Survival Strata | Treatment | Control |
| 1. Live with treatment – Live with control (LL) | a – QOL | e – QOL |
| 2. Live with treatment – Die with control (LD) | b – QOL | f – no QOL |
| 3. Die with treatment – Live with control (DL) | c – no QOL | g – QOL |
| 4. Die with treatment – Die with control (DD) | d – no QOL | h – no QOL |

2. *Only* from the subjects in stratum 1 can we get an unambiguous estimate of the average causal effect of the treatment on QOL, for only in this principal stratum are the QOL data in each of the two treatment conditions generated from a random sample from those in that stratum, untainted by intervening death.
3. So far we haven't a clue how to decide whether a person who lived and got the treatment was in stratum 1 or stratum 2, or whether someone who lived and got the control was in stratum 1 or stratum 3. And, unless we can make this decision (at least on average), our insight is not fully practical.

Let us sidestep the difficulties posed by point (3) for a moment and consider a supernatural solution. Specifically, suppose a miracle occurred and some benevolent deity decided to ease our task and give us the outcome data we require.[7] Those results are summarized in Table 4.3.

We can easily expand Table 4.3 to make explicit what is happening. That has been done in Table 4.4, in which the first row of Table 4.3 has been broken into two rows: the first representing those participants who received the treatment and the second those who received the control. The rest of the columns are the same, except that some of the entries (in *italics*) are counterfactual, given the experiment – what

[7] It is unfortunate that the benevolent deity that provided the classifications in Table 4.3 didn't also provide us with the QOL scores for all experiment participants, but that's how miracles typically work, and we shouldn't be ungrateful. History is rife with partial miracles. Why mess around with seven plagues, when God could just as easily have transported the Children of Israel directly to the Land of Milk and Honey? And while we're at it, why shepherd them to the one area in the Middle East that had no oil? Would it have hurt to whisk them instead directly to Maui?

TABLE 4.3. *True (but Partially Unobserved) Outcomes*

| Percent of Sample | Survival Stratum | Treatment Group: Live or Die | Treatment Group: QOL | Control Group: Live or Die | Control Group: QOL | Causal Effect on QOL |
|---|---|---|---|---|---|---|
| 20 | LL | Live | 900 | Live | 700 | 200 |
| 40 | LD | Live | 600 | Die | * | * |
| 20 | DL | Die | * | Live | 800 | * |
| 20 | DD | Die | * | Die | * | * |

TABLE 4.4. *True (but Partially Unobserved) Outcomes Split by Treatment Assignment*

| % of Sample | Survival Stratum | Assignment | Treatment: Survive? | Treatment: QOL | Control: Survive? | Control: QOL | Causal Effect |
|---|---|---|---|---|---|---|---|
| 10 | LL | Treatment | Live | 900 | Live | *700* | 200 |
| 10 | LL | Control | Live | *900* | Live | 700 | 200 |
| 20 | LD | Treatment | Live | 600 | Die | * | * |
| 20 | LD | Control | Live | *600* | Die | * | * |
| 10 | DL | Treatment | Die | * | Live | *800* | * |
| 10 | DL | Control | Die | * | Live | 800 | * |
| 10 | DD | Treatment | Die | * | Die | * | * |
| 10 | DD | Control | Die | * | Die | * | * |

*would* have happened had they been placed in the other condition instead. It also emphasizes that it is only from the two potential outcomes in stratum 1 that we can obtain estimates of the size of the causal effect.

This expansion makes it clearer which strata provide us with a defined estimate of the causal effect of the treatment (LL) and which do not. It also shows us why the unbalanced inclusion of QOL outcomes from the LD and DL strata misled us about the causal advantage of the control condition on QOL.

> In the affairs of life, it is impossible for us to count on miracles or to take them into consideration at all. . . .
>
> *Immanuel Kant (1724–1804)*

The combination of Rubin's Model (to clarify our thinking) and a minor miracle (to provide the information we needed to implement that clarity of thought) has guided us to the correct causal conclusions. Unfortunately, as Kant (1960) so clearly pointed out, that although miracles might have occurred regularly in the distant past, they are pretty rare nowadays, and so we must look elsewhere for solutions to our contemporary problems.

Without the benefit of an occasional handy miracle, how are we to form the stratification that allowed us to estimate the causal effect of the treatment? Here enters Paul Holland's aphorism. If we were canny when we designed this experiment, we recognized the possibility that something might happen and so gathered extra information (covariates) that could help us if something disturbed the randomization.

How plausible is it to have such ancillary information? Let us consider some hypothetical, yet familiar, kinds of conversations between a physician and a patient that might occur in the course of treatment.

1. "Fred, aside from your heart, you're in great shape. While I don't think you are in any immediate danger, the surgery will improve your life substantially." (LL)
2. "Fred, you're in trouble, and without the surgery we don't hold out much hope. But you're a good candidate for the surgery, and with it we think you'll do very well." (LD)
3. "Fred, for the specific reasons we've discussed, I don't think you can survive surgery, but you're in good enough shape so that with medication, diet, and rest we can prepare you so that sometime in the future you will be strong enough for the surgery." (DL)
4. (To Fred's family) "Fred is in very poor shape, and nothing we can do will help. We don't think he can survive surgery, and without it, it is just a matter of time. I'm sorry." (DD)

Obviously, each of these conversations involved a prediction, made by the physician, of what is likely to happen with and without the surgery. Such a prediction would be based on previous cases in which various measures of the patient's health were used to estimate the likelihood of survival. The existence of such a system of predictors allows us, without

TABLE 4.5. *Table 4.4 but now Including the Critical Covariate "Initial QOL"*

| Initial QOL | % of Sample | Survival Stratum | Assignment | Treatment Survive? | Treatment QOL | Control Survive? | Control QOL | Causal Effect |
|---|---|---|---|---|---|---|---|---|
| 800 | 10 | LL | Treatment | Live | 900 | Live | *700* | 200 |
| 800 | 10 | LL | Control | Live | *900* | Live | 700 | 200 |
| 500 | 20 | LD | Treatment | Live | 600 | Die | * | * |
| 500 | 20 | LD | Control | Live | *600* | Die | * | * |
| 900 | 10 | DL | Treatment | Die | * | Live | *800* | * |
| 900 | 10 | DL | Control | Die | * | Live | 800 | * |
| 300 | 10 | DD | Treatment | Die | * | Die | * | * |
| 300 | 10 | DD | Control | Die | * | Die | * | * |

the benefit of anything supernatural, to estimate each participant's survival stratum.

Let us use a QOL measure of each participant's health, taken as the experiment began, as a simple preassignment predictor of survival stratum. We obtain this before any opportunity for postassignment dropouts.

Table 4.5 differs from Table 4.4 in two critical ways. First, it includes the initial value of QOL for each participant; and second, the determination of who was included in each of the survival strata was based on how the value of the covariate relates to the intermediate (live/die) outcome, and not on the whim of some benevolent deity. More specifically, individual participants were grouped and then stratified by the value of their initial QOL score – no miracle required.

Those with very low QOL (300) were in remarkably poor condition, and none of them survived regardless of whether they received the treatment. Those with high QOL (800) survived whether or not they received the treatment. Those with intermediate QOL scores (500) were not in very good shape and only survived if they had the treatment, whereas they perished if they did not. And last, those with very high QOL (900) pose something of a puzzle. If they were not treated, their QOL declined to 800, which is still not bad; but if treated, they died. This is an unexpected outcome and requires follow-up interviews with their physicians and family to try to determine possible reasons for their deaths – perhaps they felt so good after the treatment that they engaged in some too-strenuous activity that turned out to be fatally unwise.

## Summary

In the past a theory could get by on its beauty; in the modern world a successful theory has to work for a living.

1. Before assignment in an experiment each subject has two potential outcomes – what the outcome would be under the treatment and what would it be under the control.
2. The causal effect of the treatment (relative to the control) is the difference between these two potential outcomes.
3. The fly in the ointment is that we only get to observe one of these.
4. We get around this by estimating the average causal effect, and through random assignment we make credible the assumption that what we observed among those in the control group is what we would have observed in the treatment group had they not gotten the treatment.
5. When someone dies (or moves away or wins the lottery or otherwise stops coming for appointments) before we can measure the dependent variable (QOL), we are out of luck. Thus, the only people who can provide estimates of the causal effect are those who would have lived under the treatment AND who would have lived under the control. No others can provide the information required.
6. So subjects who would survive under the treatment but not the control are no help; likewise those who would die under the treatment but not the control; and, of course, those who would die under both conditions add nothing.
7. So we can get an estimate of the causal effect of the treatment only from those who survive under both conditions.
8. But we don't get to observe whether they will survive under both – only under the condition to which they were assigned.
9. To determine whether they would have survived under the counterfactual condition, we need to predict that outcome using additional (covariate) information.

10. If such information doesn't exist, or if it isn't good enough to make a sufficiently accurate prediction, we are stuck. And no amount of yelling and screaming will change that.

## Conclusion

In this chapter we have taken a step into the real world, where even randomized experiments can be tricky to analyze correctly. Although we illustrated this with death as the very dramatic reason for nonresponse, the situation we sketched occurs frequently in less dire circumstances. For example, a parallel situation occurs in randomized medical trials comparing a new drug with a placebo, where if the patient deteriorates rapidly, rescue therapy (a standard, already approved drug) is used. As we demonstrated here, and in that context as well, one should compare the new drug with the placebo on the subset of patients who wouldn't need rescue, whether assigned the treatment or the placebo.

These are but a tiny sampling of the many situations in which the clear thinking afforded by Rubin's Model with its emphasis on potential outcomes can help us avoid being led astray. We also have emphasized the crucial importance of covariate information, which provided the pathway to forming the stratification that revealed the correct way to estimate the true value of the causal effect. For cogent and concise description, we have used a remarkably good covariate (initial QOL), which made the stratification of individuals unambiguous. Such covariates are welcome but rare. When such collateral information is not quite so good, we will need to use additional techniques, whose complexity does not suit the goals of this chapter.

What we have learned in this exercise is that:

(1) The naïve estimate of the causal effect of the treatment on QOL, derived in Table 4.1, was wrong.
(2) The only way to get an unambiguous estimate of the causal effect was by classifying the study participants by their potential outcomes, for it is only in the survival stratum occupied by participants who would have lived under either treatment or control that such an unambiguous estimate can be obtained.

(3) Determining the survival stratum of each participant can only be done with ancillary information (here their prestudy QOL). The weaker the connection between survival stratum and this ancillary information, the greater the uncertainty in the estimate of the size of the causal effect.

This was a simplified and contrived example, but the deep ideas contained are correct, and so nothing that was said needs to be relearned by those who choose to go further.

# 5

# Using Experiments to Answer Four Vexing Questions

*Quid gratis asseritur, gratis negatur.*[1]

## Introduction

In Chapter 3 we discussed how an observed correlation between high self-esteem (happiness) and school performance gave rise to various causal theories, at least one of which, if acted upon, could lead to unhappy outcomes. We saw how a simple experiment, by measuring the size of the causal effect, could provide evidence that would allow us to sift through various claims and determine the extent to which each was valid or specious. Every day we encounter many correlations that some claim to indicate causation. For example, there is a very high correlation between the consumption of ice cream and the number of drownings. Some cautious folk, seeing such a strong relation, have suggested that eating ice cream should be sharply limited, especially for children. Happily, cooler heads prevailed and pointed out that the correlation was caused by a third variable, the extent of warm weather. When the weather is warm, more people go swimming, and hence risk drowning, and more people consume ice cream. Despite the high correlation, neither eating ice cream nor swimming tragedies are likely to cause the weather to warm

[1] This is an ancient Latin proverb translated by Christopher Hitchens as, "What can be asserted without evidence can also be dismissed without evidence." The translation appeared in his 2007 book *God Is Not Great: How Religion Poisons Everything* (p. 150). New York: Twelve Books.

up. We could design an experiment to confirm this conjecture, but it hardly seems necessary.

After it was noted that infants who sleep with the lights on were more likely to be nearsighted when they grew up, advice to new parents was both rampant and adamant to be sure to turn the lights off when their baby went to sleep. Only later was it revealed that nearsightedness has a strong genetic component and nearsighted parents were more likely to leave lights on. Again, a controlled experiment – even a small one, though practically difficult to do – would have made clear the direction and size of the causal connection.

As one last example, consider the well-known fact that men who are married live longer, on average, than those who are single. The usual causal inference is that the love of a good woman, and the regularity of life that it yields, is the cause of the observed longer life.[2] In fact, the causal arrow might go in the other direction. Men who are attractive to women typically are those who are healthier and wealthier, and hence more likely to live longer. Men who are in poor health and/or of limited means and prospects find it more difficult to find a wife.

All of these examples exhibit the confusion that often accompanies the drawing of causal conclusions from observational data. The likelihood of such confusion is not diminished by increasing the amount of data, although the publicity given to "big data" would have us believe so. Obviously the flawed causal connection between drowning and eating ice cream does not diminish if we increase the number of cases from a few dozen to a few million. The amateur carpenter's complaint that "this board is too short, and even though I've cut it four more times, it is still too short," seems eerily appropriate.

It is too easy for the sheer mass of big data to overwhelm the sorts of healthy skepticism that is required to defeat deception. And now, with so much of our lives being tied up with giant electronic memory systems, it is almost trivial for terabytes of data to accumulate on almost any topic. When sixteen gazillion data points stand behind some hypothesis, how could we go wrong?

[2] At least one wag has suggested that life for married men is not longer; it just seems that way.

Let me make a distinction at this point between data and evidence. Data can be anything you have: shoe sizes, bank balances, horsepower, test scores, hair curl, skin reflectivity, family size, and on and on. Evidence is much narrower. Evidence is data related to a claim. Evidence has two important characteristics, both of which are important in determining its value in supporting or refuting the claim: (1) how related is the evidence to the claim (the *validity* of the evidence), and (2) how much evidence is there (the *reliability* of the evidence).

On March 5, 2015, Jane Hsu, the principal of Public School 116 on Manhattan's east side announced that henceforth they would ban homework for students in fifth grade or younger. She blamed the "well-established" negative effects of homework on student learning for the policy change. What would be evidence that would support such a claim? It is easy to construct an argument to support both sides of this issue. I'm sure Ms. Hsu has been the recipient of phone calls from parents complaining about the limited time their children have for other afterschool activities, and how stressed out both they and their children are trying to fit in the of learning multiplication tables, soccer practice, and ballet lessons. On the other side of the argument it is hard to imagine a basketball coach banning her charges from practicing foul shots or a piano teacher not assigning the practicing of scales (to say nothing about teachers asking their students to practice reading and, yes, learning math facts). Amassing evidence in support of, or refuting, the claims about homework for young children having a negative effect is just building an argument. Both aspects of the evidence determine the strength of that argument: their validity and their reliability.

Looking at the number of Google hits on travel and entertainment websites from people who lived in PS 116's district would be data, but not evidence; and their evidentiary value would not improve even if there were millions of hits.

The mindless gathering of truckloads of data is mooted by the gathering of even a small amount of thoughtfully collected evidence. Imagine a small experiment in which we chose a few second grade classrooms from PS 116 and randomly assigned half of each class to read something at home for, say, an hour a week, while allowing the other half to do whatever they wanted. Then after a semester we compared the gains in the reading scores of the two groups from the beginning of the school year. If we found that the group assigned to read showed greater gains we would have evidence that their homework made a difference. If similar studies were done assigning the learning of math facts and found similar gains this would add credence to the claim that homework helps for those topics for that age child. If the same experiment was repeated in many schools and the outcome was replicated, the support would be strengthened. Of course, this experiment does not speak to concerns of homework causing increased stress among students and parents, but that too could be measured.

My point, and the point of the rest of this chapter (as well as most of this book), is that for causal claims a small, thoughtfully designed experiment can actually provide a credible answer that would elude enormous, but undirected, data gathering.[3]

## My Dream

I have recurrent dreams in which I am involved in a trial on which I have shifted from my usual role as a witness and instead have been thrust onto the bench in judge's robes. One such dream involved an examinee with a disability who had requested an accommodation from the testing organization. After a careful study of the supporting materials submitted by the examinee she was granted a 50 percent increase in testing time. The examinee did not feel that was sufficient and petitioned for more, but was turned down. The final step in the appellate process was the hearing in which I was presiding. The examinee argued that because of her physical limitations she required, at a minimum, double the standard time. The testing company disagreed and said that, in their expert opinion, 50 percent more time was sufficient.

The goal of accommodations on exams is to level the field so that someone with some sort of physical disability is not unfairly disadvantaged. But balance is important. We do not want to give such a generous accommodation that the examinee with a disability has an unfair

[3] *Causal* claims are ill tested with *casual* data gathering, despite the superficial resemblance of the two words (differing only by a vowel movement).

advantage. Both sides agreed that this was not a qualitative question; the examinee ought to have some accommodation. The issue, in this instance, is the quantitative question "how much is enough?"[4]

As I sat there on the bench thinking about the two arguments, I felt the heavy weight of responsibility. Surprisingly, my mind shifted back more than a century to a conversation between Sherlock Holmes and John Watson. Watson asked the great detective what he made of the scant clues before them, and Holmes replied, "Watson, you know my methods. What do you think?"

Over the past forty years it has been my good fortune to work with two modern masters, Paul Holland and Don Rubin. Though I am no more a substitute for them than Watson was for Holmes, I know their methods.

And so sitting there, very much alone on that bench, I tried to apply those methods.

The key question was one of causal inference – in this instance the "treatment" is the size of the accommodation, and the causal effect of interest is the extent of the boost in score associated with a specified increase in testing time. The effect, once measured, then has to be compared with how much of a boost is enough. With the question phrased in this way, the character of a study constructed to begin to answer it was straightforward: randomly assign examinees to one of several time allotments and look at the relation between time and score. The function connecting accommodation length to score could then inform decisions on the size of the accommodation. It would also tell us how important it is to get it exactly right. For example, if we find that the connecting function is flat from 50 percent more time onward, the testing organization could allow as much time as requested without compromising fairness or validity of the test.

This chain of reasoning led me to ask the representatives of the testing organization, "What is the relation between testing time and score?" They said that they didn't know exactly, but were pretty sure that it was monotonically increasing – more time meant higher scores.

[4] This, in its most general interpretation, is among the most important existential questions of our age, but I will leave an exploration of this great question to other accounts.

I thought to myself, "That's a start, but not an answer. Yet it takes a while to see the need for such studies, to design them, and to carry them out. Perhaps they haven't had enough time to get them done." This led to my second question, "How long have you been giving this test with extra time as an accommodation?"

They replied, "About fifteen years."

I nodded and said, "That's enough time to have done the studies necessary to know the answer, yet you haven't done so. I find for the plaintiff."

It is too bad that real life is so rarely as satisfying as fantasy.

## On the Role of Experiments in Answering Causal Questions

In medical research a random-assignment controlled experiment has long been touted as the "gold standard" of evidence. Random assignment is a way to make credible the crucial assumption that the experimental and control groups were the same (on average) on all other variables, both measured and unmeasured. Without such an assumption we cannot be sure how much of the posttreatment result observed was due to the treatment and how much was due to some inherent difference between the groups that we did not account for.

Of course random assignment is not always practical, and for these cases there is a rich literature on how to make the assumption of no difference between experimental and control groups plausible in an observational study.[5] But even if done perfectly, an observational study can only approach, but never reach, the credibility of randomization in assuring that there is no missing third variable that accounts for the differences observed in the experimental outcome.

Despite the acknowledged power of this methodology, it is too rarely used in education. But when it is used correctly on an important question,

[5] In an *experimental* study the people running the study determine all aspects it: what is the treatment, what is the control, who gets which, and what is the dependent variable. In an *observational* study the control of one or more of these aspects is lacking. Most commonly those in the treatment group self-select to do it (e.g., to measure the impact of smoking on health, we might compare smokers with nonsmokers, but people are not assigned to smoke, they choose to smoke or not).

Rosenbaum 2002 and 2009 are the standard references in the field.

it can provide us with an answer. That is one reason why the Tennessee study of class size in the early school grades (discussed in many places – Mosteller 1995 gives an especially clear and comprehensive description) has assumed such a (well-deserved) place of prominence in educational research; it has provided a credible answer.

In the interest of expanding the use of this powerful methodology, let me give four very different examples of how it can be used. In two of these it already has begun to be applied, but this should not deter researchers from doing more of it in the same problem area; there is plenty of darkness to go around.

## Problem 1: Accommodation for Examinees with Disabilities

The experiment that I imagined in the dream that began this chapter lays out the basic idea. It is closely akin to the method of low-dose extrapolation that is common in the study of drug efficacy and widely used in so-called Delaney Clause research. That clause in the law forbids any food additive that is carcinogenic. The challenge is to construct a dose-response curve for an additive suspected of being a possible carcinogen. To accomplish this, one assigns the experimental animals randomly to several groups, gives each group a different dosage of the additive and then records the number of tumors as a function of the dose. But, and here is the big idea, for most additives, tumors are rare with the dosages typically encountered in everyday life. So, to boost the size of the effect, one daily dose might be, say, the equivalent amount of artificial sweetener found in twenty cases of diet soda; a second group might have that contained in ten cases/day; and a third might be five cases/day. Such massive doses, though unrealistic, would accentuate any carcinogenic effect the additive might have. To the results of this experiment we fit a function connecting dose and response and extrapolate downward to the anticipated low doses, hence the name *low-dose extrapolation.*

This methodology can help us provide necessary accommodations fairly. The experiment would take a sample of examinees without any disabilities and randomly assign them to a number of different time limits, say, normal time, 25 percent more than normal, 50 percent more, 100 percent, 200 percent, and unlimited. Then administer the test and keep track of the mean (or median) score for each group and connect the results with some sort of continuous interpolating function. With this function in hand we can allow examinees to take the test with whatever amount of time they prefer and estimate what their score would be with unlimited time (or, if required, what it would be for any specific time). Of course, it is not credible that the same function would suit examinees with disabilities, but we can allow examinees who require a time accommodation to have unlimited time. We can then compare their scores with the estimated asymptotic scores of the rest of the population. In this way we can make fair comparisons among all examinees, and thus there is no need to flag any scores as having been taken under nonstandard conditions.

Some practical questions may need to be resolved, but this approach provides the bones of an empirical solution to a very difficult problem.[6]

Note the importance of the experimental design. Had we instead used an observational approach and merely kept track of how long each examinee took, and what their score was, we would likely find that those who finished most quickly had the highest scores and those who took the most time had the lowest. We would then reach the (unlikely) conclusion that to boost examinees' scores we should give them less time. Reasoning like this led researcher Roy Freedle in 2003 to conclude that we could reduce the race differences in standardized tests by making them more difficult; an approach that was subsequently dubbed Freedle's Folly to commemorate the efficacy of his method.[7]

An experiment in which examinees were randomly assigned to different time allocations was performed on an October 2000 administration of the SAT.[8] The practical limitations involved in separately timing a specific section of the SAT led to a clever alternative. Instead, a section of the verbal portion of the test (which is not part of the examinee's score), which had twenty-five items on it, was modified so that some examinees had two more items added in randomly, others had five more, and a fourth group had ten more items. But only the core twenty-five were

[6] For more details see Wainer 2000 and chapter 7 in Wainer 2009.
[7] Chapter 8 in Wainer 2009.
[8] Wainer et al. 2004; Bridgeman, Trapani, and Curley 2004.

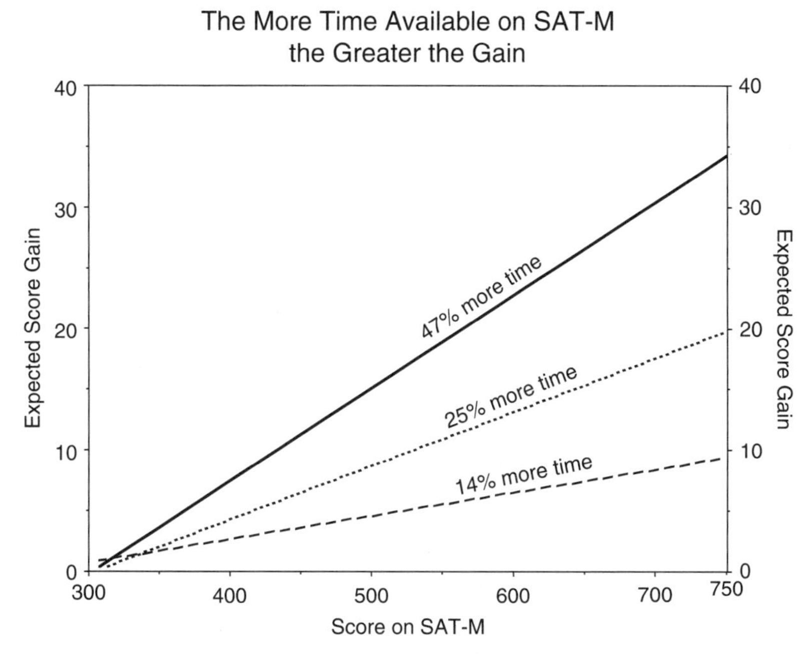


FIGURE 5.1. Expected gain scores on the various experimental math sections over what was expected from the standard-length section. Results are shown conditional on the score from the appropriate operational SAT-M form.

scored. Obviously, those with more items had less time per item. A parallel manipulation was done on the math portion of the exam. What they discovered was revealing and important. On the math portion of the SAT the more time allocated, the higher the score, but not uniformly across the score distribution (see Figure 5.1). For those who had an SAT-M score of 300, extra time was of no benefit. For those with a score of 700, 50 percent more time yielded a gain of almost forty points. The inference was that examinees of higher ability could sometimes work out the answers to hard questions, if they had enough time, whereas examinees of lower ability were stumped, and it didn't matter how much time they had.

The results from the verbal portion of the exam (see Figure 5.2) tell a different, but equally important, story. It turned out that extra time, over the time ordinarily allocated, had almost no effect. This means that extra time can be allocated freely without concern that it will confer an unfair advantage.

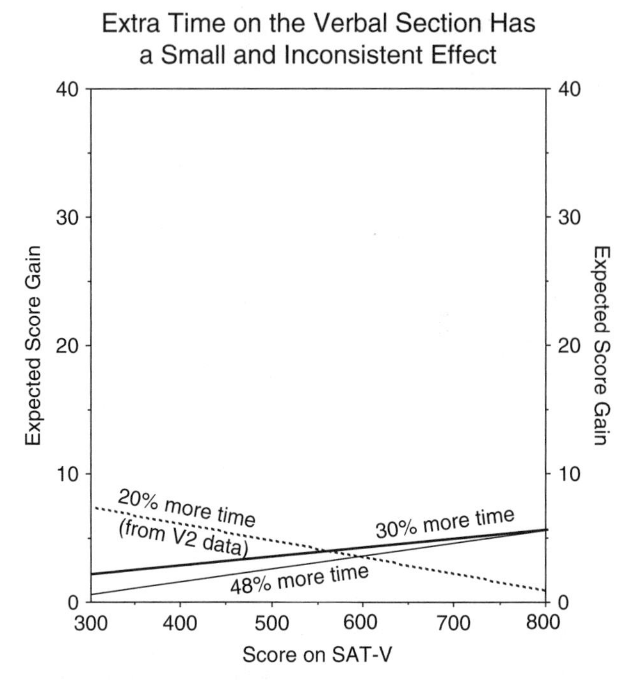


FIGURE 5.2. Allocating extra time on the SAT-V does not seem to have any consistent or substantial effect.

Of course, such studies need not be limited to setting the proper parameters for test accommodations. They can also be used to study the extent to which a test is speeded,[9] and, more quantitatively, how much an effect the speededness has on performance.[10]

## Problem 2: Unplanned Interruptions in Testing

To be comparable, test scores must be obtained under identical conditions, or if not exactly identical, as close as possible. And when conditions are not identical, we should try to measure how much change in testing conditions can be endured without affecting the score. Sometimes during the course of a test, events occur that interrupt testing. Perhaps

[9] A speeded test is one where a sizable proportion of the examinees do not have enough time to answer 95 percent of the items.

[10] It is toward the study of such aspects that, in 2004, led ETS researcher Brent Bridgeman and his colleagues in their study of the GRE as well as some as yet unpublished work done by Brian Clauser and his colleagues on a medical licensing exam.

one school had a fire scare, or a storm and the electricity went out, or an examinee got sick and disrupted the exam temporarily. When this happens, what do we do? A variety of ad hoc solutions have been practiced over the years, and perhaps they sufficed. But now, with the widespread use of computerized test administrations, the likelihood of interruptions has increased. This increase has at least two reasons: (1) computers are more complicated than #2 pencils, so there is more room for something to go wrong, and (2) computerized tests are usually administered continuously, so that there is a greater chance of something going amiss in the vastly increased time span during which the tests are being given.

All too commonly a test administration is interrupted, sometimes for just a few minutes, sometimes for a day or more.[11] What do we do when this happens? Typically, attempts to measure the effect of the interruption are made after the fact.[12] Such studies typically compare the average score before the interruption with that afterward. If the scores are the same, the testing organization heaves a sigh of relief and infers that the interruption was of no consequence and did not affect the scores' validity.

Of course, if the difficulty of the test is not the same in these two parts, that has to be adjusted for. Similarly, if there are fatigue or speededness effects, they too must be accounted for. All of these adjustments require assumptions, and all entail some error. Is such an approach good enough?

This question is fundamentally an ethical one. It is in the testing company's best interest, when there is an interruption, to look for an effect and not find one, that is, to accept the null hypothesis. To do this is easy: simply do a poor study with a small sample size, large errors, and weak statistical methods. Thus, in situations like these, it is a moral imperative to use the largest samples and the most powerful and sensitive designs possible. Then, if no significant effects are found, it adds credibility to the desired conclusion.

This sort of opportunistic before and after analysis has too many weaknesses to be acceptable if something better is possible. One approach, a definite improvement, compares the difference in the before

[11] Davis 2013; Solochek 2011; and Moore 2010.

[12] Bynum, Hoffman, and Swain 2013; Hill 2013; Mee, Clauser, and Harik 2003; and Thacker 2013.

and after scores for those who have suffered an interruption (the treatment group) with those who have not – a comparison group. Because the two groups were not created by random assignment (although the interruption was unpredicted), the comparison group must be matched to the treatment group on the basis of as many suitable "covariates" as have been recorded, such as sex, ethnicity, age, and education.

Such careful observational studies of the effect of interruptions are a definite improvement over the raw change score studies, but they are too rare. In 2014, distinguished California researcher Sandip Sinharay laid out a variety of ways of constructing suitable comparison groups in observational studies to allow more credible estimates of the effects of interruptions. He also describes when such methods are not likely to be suitable because of the uncertain availability of appropriate matching groups.

There are limits to the accuracy of any reactive studies of the causal effect of an interruption. Any such study that tries to estimate causal effects suffers from the very real possibility of some missing third variable either being the actual cause of the differences observed or of mitigating a true effect by imparting another effect in the opposite direction. The most credible estimates of the size of the causal effect of an interruption come from a true experiment with random assignment of individuals to treatment groups. Running such experiments means that we must be proactive. One such study design would divide up examinees into groups defined by when and for how long interruptions occur. This approach would allow us to estimate the effect of the interruption and parameterize its effect by its location and length. Once such studies are complete, we are ready to adjust for any interruptions that subsequently occur. I don't know of anyone who has yet ventured to do such a study.

## Problem 3: Measuring the Effect of Feedback to Judges in Standard Setting

Standard setting is an exercise regularly done for almost all large-scale tests. For licensing tests it establishes the cut-score for passing; for many educational tests it establishes boundaries for the various categories of performance. Setting standards for the various NAEP[13] tests has led to the canonization

[13] National Assessment of Educational Progress.

(reification?) of the four category labels Advanced, Proficient, Basic, and Below Basic.[14] Now almost all K–12 state tests have these same categories, which are then used as if they had real meaning. Children are promoted (or not) on the basis of which category they are placed in; educational programs are judged by their success in getting students placed in the higher categories; and the careers of teachers and other school officials hang in the balance. The boundaries for these categories have to be set. To be legally defensible in the United States, these boundaries must be decided upon by the judgment of experts using one or more of several kinds of tasks.[15] Many of the most widely used methods are augmented by a feedback stage in the standard setting process. One way that this takes place is that after the judges arrive at a consensus on where the pass-fail line should be drawn (or, on the boundaries for multiple performance categories), they are told what would have been the consequences of such boundaries had they been used in a previous test administration. They might be told how, using their standards, only 20 percent of the examinees would pass, only 12 percent of minority examinees would pass, and so forth. The judges are then encouraged to reconvene and decide whether they wish to amend the boundaries they previously agreed upon. Usually the committee of experts, using their judgment in combination with the real-world information given them by those running the standard setting, iterates toward a stable, accurate, and acceptable result.

For example, suppose the standards committee goes through one of the detailed standard setting procedures and arrives at a passing score. They then find out that, had that cut-score been in use the previous year, only 30 percent of examinees would have passed instead of the more usual 70 percent. They then reconsider and decide that perhaps they had been too severe, and, with the wisdom borne of further discussion, revise the cut-score and find, with only a modest change, that 69.4 percent would now have passed. They then agree that the new cut-score is the correct one, and they go home, assured of a job well done.

Yet, anyone with either experience in doing social psychological research or a cynical nature would immediately ask: What is the causal effect of the

[14] I suspect that Below Basic was adopted as a term to avoid using the pejorative, but perhaps more evocative, label "Incompetent."

[15] See Cizek and Bunch 2007 or Zieky, Perie, and Livingston 2008 for careful and detailed discussions of many of these methods.

feedback? Does it improve judgments through its infusion of reality? Or do the experts slavishly follow wherever the feedback leads them? If the latter, we ought to either dispense with the expert panels altogether and just set the cut-scores to yield the same proportion of passing scores this year as last. Or, alternatively, do away with the feedback and accept the more error-laden, but untainted, judgments of the unaided experts.

How can we tell? If we just follow protocol, all we can tell is how much good information affects judgments. We don't know how much bad information would affect them. Can the expert ratings be manipulated anyway we want? The best, and perhaps the only, way to find out is to separate the two parts of the "treatment": the feedback and the accuracy of the feedback. This can be done in the context of an experimental design in which some judges receive accurate feedback and others receive wildly inaccurate information. The goal is to find out how much the judges' ratings can be manipulated.

Happily, in 2009 Brian Clauser and his colleagues at the National Board of Medical Examiners published one such study. They found that judges' ratings can be manipulated substantially, whether the feedback was accurate or not. This finding, if replicated, requires a serious rethinking of what standard setting tells us, how it should be carried out, and how much we can believe what comes out of it.

## Problem 4: What Is the Value of "Teaching to the Test"?

Considerable discussion has focused on the increased consequences of students' test scores for both students and school faculty. Specifically, it is claimed that too often teachers do not teach the curriculum and instead focus undue attention on test preparation. It would be worthwhile to know how much advantage "teaching to the test" conveys. An old adage says that one should not learn "the tricks of the trade" but instead should learn the trade. In the same way, perhaps students do better on exams if their instruction focuses primarily on the subject and deals only minimally with the specifics of the testing situation (e.g., practice with timing and filling in score sheet bubbles). A well-designed and well-executed experiment that explores this would be worthwhile. If such an experiment shows that extended test preparation confers no advantage, or even a disadvantage, teachers could feel comfortable staying with the curriculum and letting the test scores take care of themselves. Or, if it shows that such excessive preparation does indeed help, it would suggest restructuring the exams so that this is no longer true. One suitable experiment could have two conditions: only minimal test preparation and a heavy focus on such preparation. Students and teachers would be randomly assigned to each condition (with careful supervision to ensure that the treatment assigned was actually carried out). Then the exam scores would be compared.

I don't know of any experiments that have ever been carried out like this, although enormous numbers of observational studies have examined the value of coaching courses for major examinations. Typically such courses focus strictly on the test. The most studied are coaching courses for the SAT, and the overwhelming finding of independent researchers (not studies by coaching schools, which tend to be self-serving and use idiosyncratically selected samples) show that the effects of coaching are very modest. In 1983 Harvard's Rebecca DerSimonian and Nan Laird, in a careful meta-analysis of a large number of such studies, showed about a twenty-point gain on a 1,200-point scale, replicating and reinforcing ETS researchers Sam Messick and Ann Jungeblut's 1981 study. More recently, Colorado's Derek Briggs in 2001, using a completely different design and data set, showed a similar effect size. Other observational studies on coaching for the LSAT and the USMLE show even smaller effect sizes.

Such results, though strongly suggestive, have apparently not been sufficiently convincing to deter this practice. I would hope that a well-designed experiment that included such ancillary factors as student ability, SES, gender, and ethnicity could help guide future teaching and testing.

## Discussion and Conclusions

If you think doing it right is expensive, try doing it wrong.[16]

In this chapter I have described four very different, but important, research questions in educational measurement. In all four, approximate

[16] With apologies to Derek Bok, whose oft-quoted comment on the cost of education is paraphrased here.

answers, of varying credibility, can be obtained through observational studies. However, the data used in observational studies are found lying on the street, and because we have no control over them, we are never entirely sure of their meaning. So we must substitute assumptions for control. When we run a randomized experiment, in which we are in control of both what is the treatment and who receives it, the randomization process provides a credible substitute for the assumption of *ceteris paribus*.

Why is this? In all the situations I have discussed, the goal was always to estimate the causal effect of the treatment. Sometimes the treatment was more time, sometimes it was having the exam interrupted, and sometimes it was getting feedback about the choice of cut-scores; but in all cases we are interested in measuring the size of the causal effect.

In all cases, the causal effect is the difference between what happened when the experimental group received the treatment and the counterfactual event of what would have happened if that same group had received the control condition. But we do not know what would have happened had the experimental group received the control condition – that is why we call it a *counterfactual*. We can know what happened to the group when they got the control condition. We can only substitute the outcome obtained from the control group for the counterfactual associated with the experimental group if we have credible evidence that there are no differences, on average, between the experimental group and the control group. If the people in either group were selected by a process we do not understand (e.g., Why did someone elect to finish the exam quickly? Why was one exam interrupted and another not?), we have no evidence to allow us to believe that the two groups are the same on all other conditions except the treatment. Randomization largely removes all these concerns.

Because of the randomization the average outcome of the control group is equal to what the average would have been in the experimental group had they received the control. This is so because nothing is special about the units in either condition – any subject is as likely to be in one group as the other.

A full formal discussion of Rubin's Model is well beyond the goals of this chapter; interested readers are referred to Paul Holland's justly famous 1986 paper "Statistics and Causal Inference."[17] As I pointed out in Chapter 3, a key idea in Holland's paper (derived from Rubin's foundational 1974 paper) is that the estimate of the average outcome seen in the control group is the same as what would have been observed in the experimental group, had they been in the control condition. The truth of this counterfactual rests on the randomization.

Without the control offered by a true randomized experiment, we must substitute often-untenable assumptions of *ceteris paribus* for the power of homogeneity provided by the random assignment. And so I have argued for greater use of the gold standard of causal inference – randomized, controlled experiments – instead of the easier, but more limited, methods of observational studies.

It has not escaped my attention that the tasks associated with doing true designed experiments is more difficult than merely analyzing some data found lying in the street for an observational study. Sometimes such experiments seem impossible. Indeed sometimes they are, and then we must make do with observational studies. But experience has taught us that, if it is sufficiently important to get the right answer, experiments that might have been thought impossible can be done.

For example, suppose some horrible disease is killing and crippling our children. Suppose further that researchers have developed a vaccine that animal research gives us high hopes of it working. One approach is an observational study in which we give the vaccine to everyone and then compare the incidence of the horrible disease with what had happened in previous years. If the incidence was lower, we could conclude that the vaccine worked or that this was a lucky year. Obviously the evidence from such a study is weaker than if we had done a true random assignment experiment. But imagine how difficult that would be. The dependent variable is the number of children who come down with the disease – the size of the causal effect is the difference in that dependent variable between the vaccine group and the control group. The consequences of denying the treatment to the control group are profound. Yet, it was important enough to get the right answer that in 1954 an

[17] In my opinion, as well as those of many others, this paper ranks among the most important statistics papers of the twentieth century.

experiment was done to test the Salk vaccine against polio. It had more than five million dollars of direct costs ($43.6 million in 2014 dollars) and involved 1.8 million children. In one portion of the experiment 200,745 children were vaccinated and 201,229 received a placebo. These enormous sample sizes were required to show the effect size anticipated. There were eighty-two cases of polio in the treatment group and 162 cases in the placebo group. This difference was large enough to prove the value of the vaccine. I note in passing that there was enough year-to-year variation in the incidence of polio that if an uncontrolled experiment had been performed in 1931 the drop in incidence in 1932 would have incorrectly indicated that the treatment tried was a success.[18]

If society is willing to gamble the lives of its children to get the right answer, surely the cost of an experimental insertion of a testing delay is well within the range of tolerability.

The Salk experiment is not an isolated incident. In 1939 an Italian surgeon named Fieschi introduced a surgical treatment for angina that involved ligation of two arteries to improve blood flow to the heart. It worked very well indeed. In 1959 Leonard Cobb operated on seventeen patients; eight had their arteries ligated and nine got incisions in their chests, but nothing more. There was no difference between the treatment and the control group. Again we find that society was willing to pay the costs of having sham surgery in order to get the answer. Of course, with the advent of "informed consent" such sham surgery is much more difficult to do now, but the fact that it was done reflects on the size of the efficacy gap between observational studies and designed randomized, controlled experiments.

This brief side trip into medical experimentation helps to place the human costs of doing true experimentation within the area of education in perspective. None of the experiments I have proposed here, or variations on them, have consequences for their participants that are as profound as those seen in their medical counterparts. In addition, we must always weigh the costs of doing the experiment against the ongoing costs of continuing to do it wrong. That is the ethical crux of the decision.

[18] Meier 1977.

# 6

# Causal Inferences from Observational Studies: Fracking, Injection Wells, Earthquakes, and Oklahoma

## Introduction

On November 11, 1854, Henry David Thoreau observed, "Some circumstantial evidence is very strong, as when you find a trout in the milk." He was referring to an 1849 dairyman's strike in which some of the purveyors were suspected of watering down the product. Thoreau is especially relevant when we are faced with trying to estimate a causal effect, but do not have easily available the possibility of doing a suitable experiment, and so are constrained to using available data for an observational study.

In Chapters 3 and 4 we saw how the key to estimating the causal effect of some treatment was comparing what occurred under that treatment with what would have occurred without it. In Chapter 5 we showed how the structure of a randomized, controlled experiment was ideally suited for the estimation of causal effects. But such experiments are not always practical, and when that is the case we are constrained to use an observational study, with naturally occurring groups, to estimate the size of the causal effect. When we do not have randomization to balance the treatment and control groups we must rely on some sort of post hoc matching to make the equivalence of the two groups credible. Results from observational studies must rely on evidence that is circumstantial, and the strength of such evidence is what determines how credible the conclusions are.

The balance of this chapter deals with a single remarkable example, an exploration of a possibly causal connection between oil and gas exploration and earthquakes. More specifically, we will explore the consequences of the unfortunate combination of using a drilling technique called hydraulic fracturing (commonly called *fracking*) and the disposal of wastewater by the high-pressure injection of it back into the earth. I believe that the available evidence is, in Thoreau's delicious metaphor, a trout in the milk.

## Dewatering

An oil well is considered to be exhausted when the amount of oil it yields is no longer sufficient to justify the cost of its extraction. Most of Oklahoma's wells fell into this category by the 1990s because of the immense amount of wastewater that was brought up along with the diminishing amount of oil. But in the twenty-first century the combination of dewatering technologies and the rising price of oil made many of Oklahoma's abandoned wells economically viable again. The idea was to just pull up the water with the oil – about ten barrels of water for each barrel of oil. This has yielded billions of barrels of wastewater annually that has to be disposed of. The current method is to use high-pressure pumps to inject it back into the earth in wastewater wells.

## Fracking

Fracking is the process of drilling down into the earth before a high-pressure water mixture is directed at the rock to release the gas inside. Water, sand, and chemicals are injected into the rock at high pressure that allows the gas to flow out to the head of the well. This procedure has been in use for about sixty years. However, horizontal drilling is a new wrinkle introduced by 1990 that could dramatically increase the yield of the well. Horizontal drilling is a horizontal shaft added onto the vertical one, after the vertical drilling has reached the desired depth (as deep as two miles). This combination expands the region of the well substantially. The high-pressure liquid mixture injected into the well serves several purposes: it extends the fractures in the rock, adds lubrication, and carries materials (proppants) to hold the fractures open and thus extend the life of the well. Horizontal fracking is especially useful in shale formations that are not sufficiently permeable to be economically viable in a vertical well. The liquid mixture that is used in fracking is disposed of in the same way as the wastewater from dewatering wells.

## Concerns

The principal concern about the use of fracking began with the volume of water used in their operation (typically two to eight million gallons per well) and the subsequent possible contamination of drinking water if the chemicals used in fracking leached into the groundwater. But it was not too long before concerns arose about the disposal of wastewater generated from fracking and dewatering, causing a substantial increase in seismic activities. Most troubling was a vast increase in earthquakes in areas unused to them.[1] It is the concern that most of these earthquakes are manmade that is the principal focus of this chapter.

## A Possible Experiment to Study the Seismic Effects of Fracking

If we had a free hand to do whatever we wished to estimate the causal effect that the injection of large quantities of wastewater has on earthquakes, all sorts of experiments suggest themselves. One might be to choose a large number of geographic locations and pair them on the basis of a number of geological characteristics, then choose one of each pair at random and institute a program of water injection (the treatment group) and leave the other undisturbed (the control group). Of course, we would have to make sure that all the areas chosen were sufficiently far from one another that the treatment does not have an effect on a member of the control group. Then we start the experiment, keep track of the

[1] A January 2015 study in *The Bulletin of the Seismological Society of America* indicates that fracking built up subterranean pressures that repeatedly caused slippage in an existing fault as close as a half-mile beneath the wells (http://www.seismosoc.org/society/press_releases/BSSA_105-1_Skoumal_et_al_Press_Release.pdf [accessed August 27, 2015]).

number of earthquakes in the region of the treatments, and keep track of the number of earthquakes in the control regions.

It might take some time, but eventually we would have both a measure of the causal effect of such injections and a measure of the variability within each of the two groups.

While it may be pretty to contemplate such a study, it isn't likely to be done, and it certainly won't get done any time soon. Waiting for such a study before we take action is of little solace to those people, like Prague, Oklahoma, resident Sandra Ladra, who on November 5, 2011 landed in the hospital from injuries she suffered when the chimney of her house collapsed in a 5.7 magnitude earthquake (the largest ever recorded in Oklahoma) – the same series of quakes that destroyed fifteen homes in her neighborhood as well as the spire on Benedictine Hall at St. Gregory's University, in nearby Shawnee. Subsequently, researchers analyzed the data from that quake[2] and concluded that the quake that injured Ms. Ladra was likely due to injection of fluids associated with oil and gas exploration. The quake was felt in at least seventeen states but that "the tip of the initial rupture plane is within ~200m of active injection wells."

## One Consequence of Not Having Good Estimates of the Causal Effect

It isn't hard to imagine the conflicting interests associated with the finding of a causal effect associated with oil and gas exploration in Oklahoma. Randy Keller, director of the Oklahoma Geological Survey, posted a position paper saying that it believes that the increase in earthquakes is the result of natural causes. In 2014, when faced with the increase of seismic activity, Mary Fallin, the governor of Oklahoma, advised Oklahomans to buy earthquake insurance. Unfortunately, many policies specifically exclude coverage for earthquakes that are induced by human activity.

[2] Keranen et al. (June 2013).

## An Observational Study

So we are faced with the unlikely event of doing a true, random assignment; an experiment on the causal effects of the combination of fracking; a high-volume wastewater injection on seismic activity; and the urgent need to estimate what is that effect. What can we do? The answer must be an observational study. One way to design an observational study is to first consider what would be the optimal experimental design (like the one I just sketched) and try to mimic it within an observational structure.

**Treatment Condition.** Treatment condition is oil exploration using fracking and dewatering in which the wastewater generated is injected under pressure into disposal wells. This will be in the state of Oklahoma during the time period 2008 to the present, which is when these techniques became increasingly widespread.

**Control Condition.** Don't do it; no fracking and, especially, no disposal of wastewater using high-pressure injection into disposal wells. The control condition would be what existed in the state of Oklahoma for the thirty years from 1978 until 2008, and for the same time period in the state of Kansas, which abuts Oklahoma to the north. Kansas shares the same topography, climate, and geology, and, over the time period from 1973 to the present, has had far less gas and oil exploration.

**Dependent Variable.** The dependent variable is the number of earthquakes with magnitude of 3.0 or greater. We chose 3.0 because that is an earthquake that can be felt without any special seismic detection equipment. Since Oklahoma has begun to experience increased seismic activity the U.S. and Oklahoma Geological Surveys (USGS and OGS) have increased the number of detectors they have deployed in the state, and so some of the increase in detected small quakes could be a function of detection and not incidence.

## A Trout in the Milk

In Figure 6.1 (from the USGS) we see Oklahoma's seismic activity summarized over the past thirty-eight years.

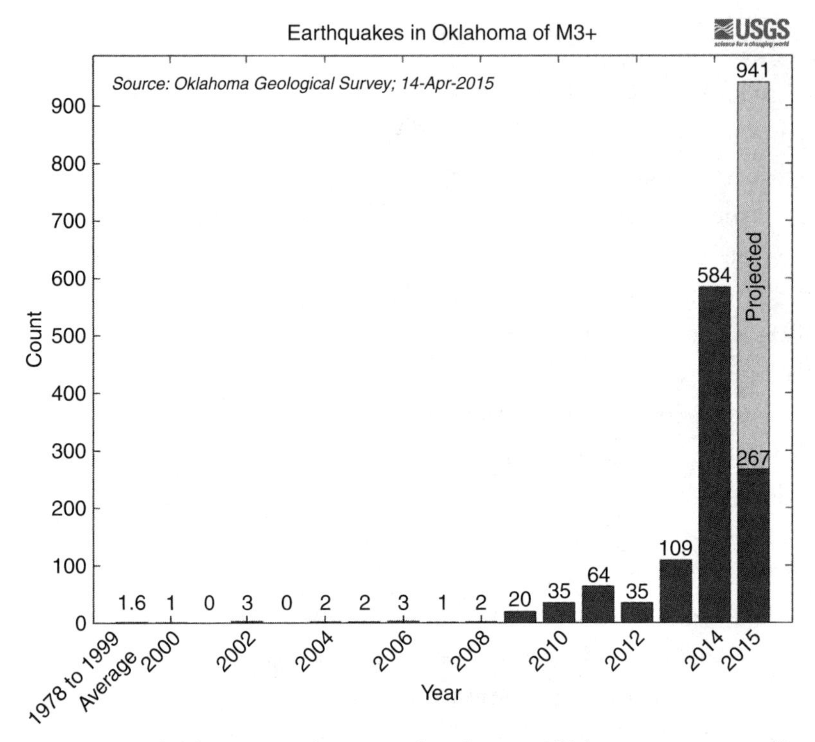


FIGURE 6.1. The frequency of 3.0+ earthquakes in Oklahoma since 1978 (from the USGS).

By the end of 2014 there had been 585 earthquakes of magnitude 3.0 or greater. If smaller earthquakes were to be included the total would greater than five thousand! So far, in 2015 there has been an average of two earthquakes of magnitude 3.0 or greater *per day*.

In the thirty years preceding the expansion of these methods of oil and gas exploration there averaged fewer than two earthquakes of magnitude 3.0 or greater *per year*.

This three-hundred-fold increase has not gone unnoticed by the general population. Oklahomans receive daily earthquake reports like they do weather. Oklahoma native, and New Yorker writer, Rivka Galchen reports that driving by an electronic billboard outside Oklahoma City last November he saw, in rotation, "an advertisement for one per cent cash back at the Thunderbird Casino, an advertisement for a Cash N

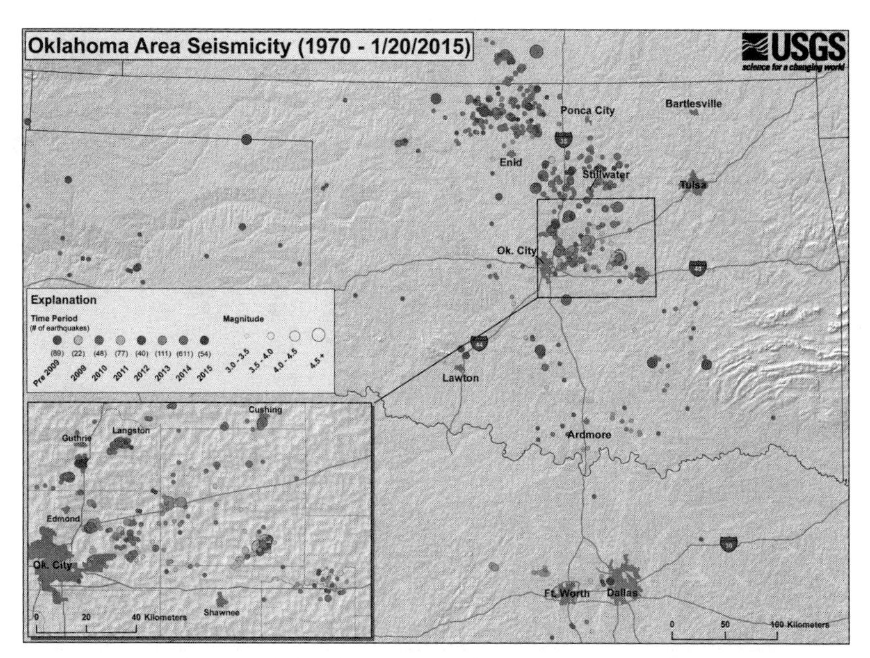


FIGURE 6.2. The geographic distribution of 3.0+ earthquakes in Oklahoma since 1970. The clusters surround injection wells for the disposal of wastewater (from the USGS). See color version facing page 108.

Gold pawnshop, a three-day weather forecast, and an announcement of a 3.0 earthquake in Noble County." Driving by the next evening he saw that "the display was the same, except that the earthquake was a 3.4 near Pawnee."

The geographic distribution of quakes is shown in Figure 6.2, also from the USGS, in which the blue dots represent the eighty-nine earthquakes in the thirty-nine years prior to 2009, the other 960 dots represent the five and a quarter years since then.

Finally, what about the control group? What was the seismic activity in Kansas during this same period? Figure 6.3 is a similar map, although the coding of the plotted points is different than in the Oklahoma map. Here the color coding reflects the depth of the seismicity and its diameter is its magnitude. The four quakes shown for the period 1973 to the present were all shallow and of 3.5 to 4.0 magnitude.

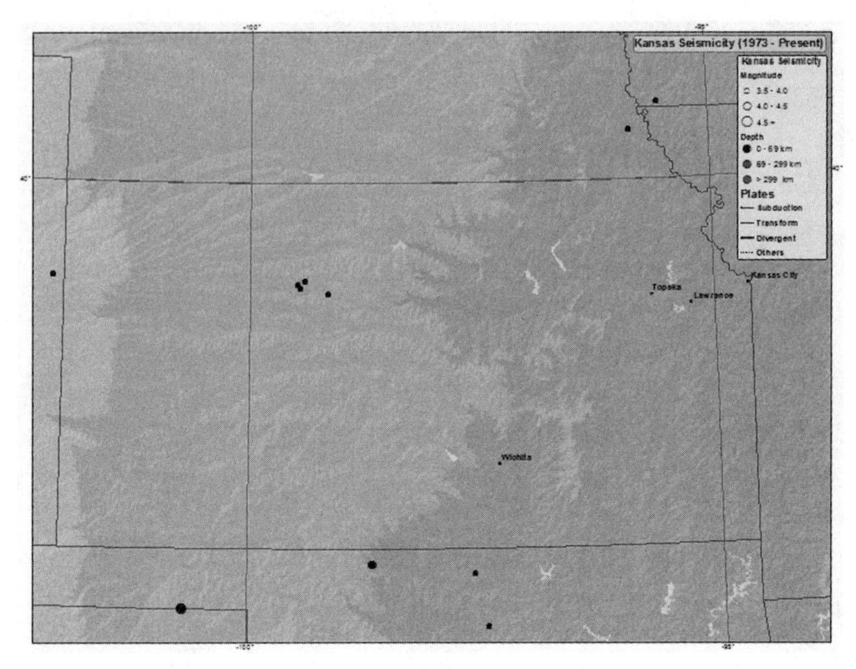


FIGURE 6.3. The geographic distribution of 3.5+ earthquakes in Kansas since 1973 (from the USGS). See color version facing page 108.

## Conclusions

The inferences that can credibly be drawn from observational studies have limitations. Consider the well-established fact that among elementary school children there is a strong positive relation between scores on reading tests and shoe size. Do large feet help you to read better? Or does reading help to stimulate the growth of one's feet? Sadly, neither. Instead, there is a third variable, age, that generates the observed relation. Older children are typically both bigger and better readers. An observational study that does not adjust for this variable would draw the wrong inference. And there is always the possibility of such a missing third variable unless there is random assignment to the treatment and control groups. For it is through randomization all missing third variables, known or unknown, are balanced on average.

The evidence presented here makes it clear that there is a strong positive relation between wastewater injection and earthquakes. But, because it is an observational study, we don't know for sure whether or not there is some missing third variable that would explain the observed phenomenon, and the size of the apparent causal connection could shrink, or even disappear.

However, no one would believe that foot size has any direct causal connection with reading proficiency, because we know about reading (and feet). Similarly, we (or at least trained geologists) know about the structure of earthquakes and the character of the rock substrata that lies beneath the state of Oklahoma, we can draw credible causal conclusions about the evidence presented here.

The inferences to be drawn from these results seem straightforward to me. I could not imagine what missing third variable might account for what we have observed. What other plausible explanation could there be for the huge increase in seismic activity? But what I believe is of small importance. The interpretations expressed by highly trained and knowledgeable geologists are much more credible. What do they think?

In an interview with Rivka Galchen, William Ellsworth, a research geologist at the USGS said, "We can say with virtual certainty that the increased seismicity in Oklahoma has to do with recent changes in the way that oil and gas are being produced.... Scientifically, it's really quite clear." There is a substantial chorus of other geologists who echo Ellsworth's views in the recent scientific literature.

But not everyone sees it that way. Michael Teague, Oklahoma's Secretary of Energy and the Environment, in an interview with the local NPR station, said "we need to learn more." His perspicacity on environmental topics was illuminated when he was asked if he believed in climate change. He replied that he believed that climate changed every day.

On April 6, 2015, CBS News reporter Manuel Bojorquez interviewed Kim Hatfield, who is with the Oklahoma Independent Petroleum Association. She said the science to prove a definitive link simply isn't there. "Coincidence is not correlation," said Hatfield. "This area has been seismically active over eons and the fact that this is unprecedented in our experience doesn't necessarily mean it hasn't happened before."

Her view was echoed by Jim Inhofe, Oklahoma's senior senator, who, in a message passed on to me by Donelle Harder, his press secretary on April 8, 2015, said that "Oklahoma is located on a fault line and has always had seismic activity. While there has been an increase in activity over the past year, the data on earthquakes that is being examined only goes back to 1978. Seismic activity is believed to have been going on for thousands of years in Oklahoma, so looking at just the last 35 years to make definitive conclusions about trends and industry connections is short sighted.... We shouldn't jump to making rash conclusions at this point. Many credible organizations, such as the National Academies, have said there is very little risk of seismic activity from the practice of hydraulic fracturing. What is being closely examined is whether wastewater disposal, which is being regulated and comes from a number of sources than just oil and gas extraction, is causing seismic activity. The scientists are looking at it, and before the issue becomes hyper-politicized by environmentalists, we need to let them do their jobs so we get reliable science."

The evidence I have presented here is certainly circumstantial, but compelling nonetheless. Despite Senator Inhofe's suggestion to the contrary, there have been a substantial number of studies published by the foremost of authorities in the most prestigious of peer-reviewed journals that support the close causal relation between fracking, and, especially, its associated wastewater disposal, to the onslaught of earthquakes that have besieged Oklahoma.[3] I have not been able to find any credible reports to the contrary.

The petroleum industry has a huge footprint in Oklahoma, and so it is easy to understand why state officials would find it hard to acknowledge evidence linking their activities to negative outcomes, regardless of the credibility of that evidence. I don't know how experimental evidence would have been greeted, but I suspect the same sorts of questions would be raised. It is reminiscent of the April 14, 1994 congressional testimony of the CEOs of the seven largest tobacco companies, who all swore that to the best of their knowledge, nicotine was not addictive.

The responses by those who deny that the dramatic increase in earthquakes is due to human activity are eerily parallel to those who deny that global warming has a human component. Indeed these are often the same people.[4]

[3] To cite three recent ones see Hand (July 4, 2014); Keranen et al. (June 2013); and Keranen et al. (July 25, 2014).

[4] In the preface we were introduced to Senator Inhofe, who took the floor in the U.S. Senate holding a snowball he made just a few minutes earlier. He indicated that this was proof positive of the fallaciousness of global warming. Of course, if we were to take seriously the role of human activity on both climate and earthquake it could profoundly affect the petroleum industry. It shouldn't be a surprise that anyone in thrall to that industry would find it difficult to be convinced of their culpability.

An argument often used in both situations (climate change and increased seismicity) is that we have had such symptoms before; we have had heat waves and droughts before, many worse than what we have now, so why credit global warming? And similarly, from Senator Inhofe, "Seismic activity is believed to have been going on for thousands of years in Oklahoma."

Both statements are undoubtedly true; is there an effective answer? Maybe. I overheard a discussion some time ago between a successful businessman and the physicist John Durso that seems relevant. The businessman argued that sure there were some hot days and some strong storms, but over the course of his life he could remember hotter days and stronger storms. He didn't buy this as evidence of global warming. Professor Durso thought for a moment and offered this analogy. "Consider Main Street in your town. People drive on it, and sometimes there is an accident. But remember a couple of years ago, when a short section of the interstate highway that goes by your town had to close down for repairs. During the repairs they detoured the traffic off the highway just before your town, routed it down Main Street, and then back on the highway afterwards. There was a substantial increase in traffic on Main Street during the repairs. There was also a substantial increase in accidents during that period. While it is certainly true that you couldn't point to any one accident and say it was caused by the closure of the highway, you would be foolish to assert that the increase in accidents was not related to the closure."

The businessman nodded in agreement. The combination of the facts and the argument convinced him. Sadly, wisdom borne of long experience tells me that he was an unusual man. The evidence I have described here is, to me at least, the very incarnation of Thoreau's fish, yet I don't delude myself into thinking that an argument made up only of logic and evidence will be enough to sway everyone. But it is a place to start.[5]

[5] But not to finish; we can gather more evidence in support of the claim that the lion's share of the seismic activity in Oklahoma is manmade by keeping track (1) of what happens when other states ignore Oklahoma's experience and institute their own programs of wastewater disposal through injection wells (e.g. North Dakota and Alberta, Canada) and (2) of seismic activity should there be a substantial decline in such wastewater disposal. The latter is not likely to yield a fast answer, for there appears to be a substantial latency possible. But those are avenues for adding to the supporting evidence.

# 7

# Life Follows Art

## *Gaming the Missing Data Algorithm*

In 1969 Bowdoin College was pathbreaking when it changed its admissions policy to make college admissions tests optional. About one-third of its accepted classes took advantage of this policy and did not submit SAT scores. I followed up on Bowdoin's class of 1999 and found that the 106 students who did not submit SAT scores did substantially worse in their first year grades at Bowdoin than did their 273 classmates who did submit SAT scores (see Figure 7.1). Would their SAT scores, had they been available to Bowdoin's admissions office, have predicted their diminished academic performance?

As it turned out, all of those students who did not submit SAT scores, actually took the test, but decided not to submit them to Bowdoin. Why? There are many plausible reasons, but one of the most likely ones was that they did not think that their test scores were high enough to be of any help in getting them into Bowdoin. Of course, under ordinary circumstances, this speculative answer is not the beginning of an investigation, but its end. The SAT scores of students who did not submit them have to be treated as missing data – at least by Bowdoin's admissions office, but not by me. Through a special data-gathering effort at the Educational Testing Service we retrieved those SAT scores and found that while the students who submitted SAT scores averaged 1323 (the sum of their verbal and quantitative scores), those who didn't submit them averaged only 1201 – more than a standard deviation lower! As it turned out, had the

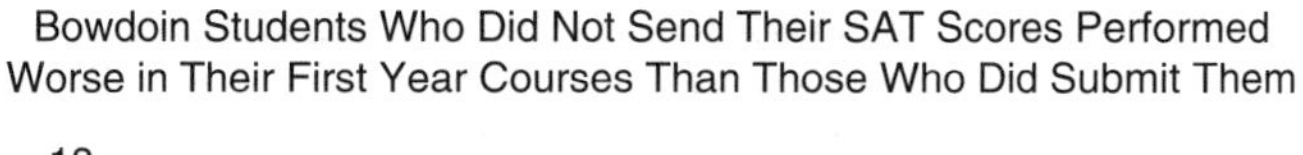


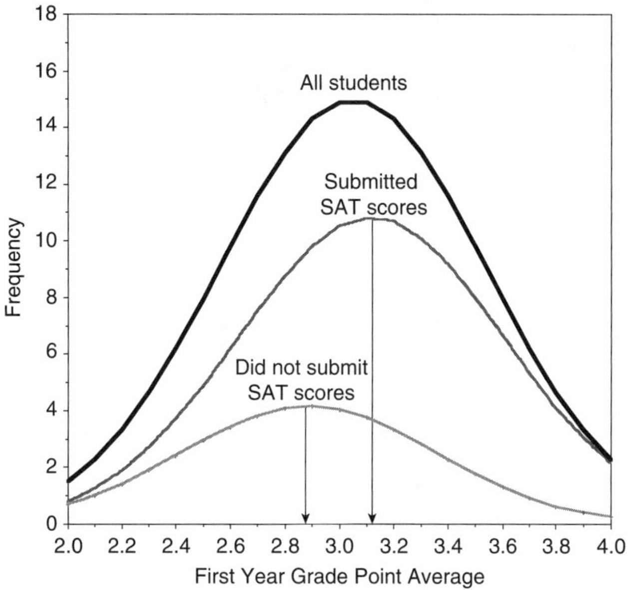


FIGURE 7.1. A normal approximation to the distributions of first-year grade point averages at Bowdoin shown as a function of whether or not they chose to submit their SAT scores.

admissions office had access to these scores they could have predicted the lower collegiate performance of these students (see Figure 7.2).

Why would a college opt for ignorance of useful information? Again there is a long list of possible reasons, and your speculations are at least as valid as mine, so I will focus on just one: the consequences of treating missing data as missing at random (that means that the average missing score is equal to the average score that was reported, or that those who did not report their SAT scores did just as well as those who did). The average SAT score for Bowdoin's class of 1999 was observed to be 1323, but the true average, including all members of the class was 1288. An average score of 1323 places Bowdoin comfortably ahead of such fine institutions as Carnegie Mellon, Barnard, and Georgia Tech, whereas 1288 drops Bowdoin below them. The influential *US News and World Report* college rankings use average SAT score as an important component. But those rankings use the reported scores as the average, essentially

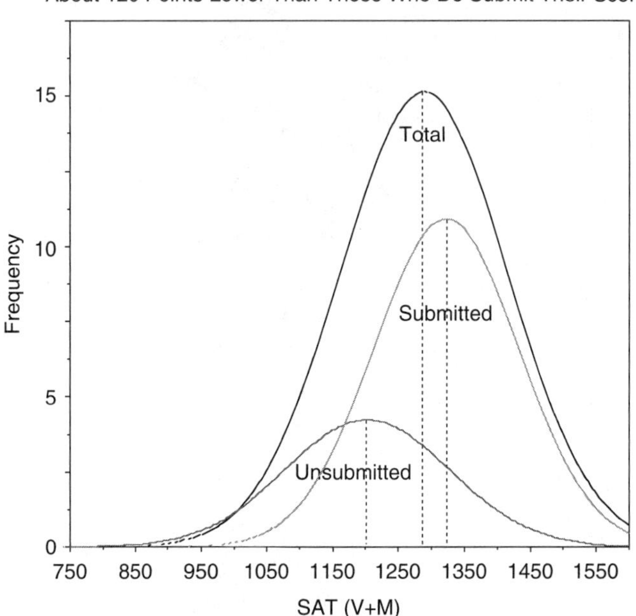


FIGURE 7.2. A normal approximation to the distributions of SAT scores among all members of the Bowdoin class of 1999.

assuming that the missing scores were missing at random. Thus, by making the SAT optional, a school could game the rankings and thereby boost their placement.

Of course, Bowdoin's decision to adopt a policy of "SAT Optional" predates the *US News and World Report* rankings, so that was almost certainly not their motivation. But that cannot be said for all other schools that have adopted such a policy in the interim. Or so I thought.

After completing the study described in the preceding text I congratulated myself on uncovering a subtle way that colleges were manipulating rankings – silly, pompous me. I suspect that one should never assume subtle, modest manipulations, if obvious large changes are so easy. *US News and World Report* gets their information reported to them directly from the schools, thus allowing the schools to report anything they damn well please.

In 2013 it was reported that six prestigious institutions admitted falsifying the information they sent to *US News and World Report* (and also the U.S. Department of Education and their own accrediting agencies).[1]

Claremont McKenna College simply sent in inflated SAT scores; Bucknell admitted they had been boosting their scores by sixteen points for years; Tulane upped theirs by thirty-five points; and Emory used the mean scores of all the students that were admitted, which included students who opted to go elsewhere – they also inflated class ranks! And there are lots more examples.

In Chapter 8 of my 2011 book *Uneducated Guesses*, I discussed the use of value-added models for the evaluation of teachers. As part of this I described the treatment of missing data that was currently in use by the proponents of this methodology. The basic idea underlying these models is to divide the change in test scores – from pretest scores at the beginning of the school year to posttest scores at the end – among the school, the student, and the teacher. The average change associated with each teacher was that teacher's "value-added." There were consequences for teachers with low value-added scores and different ones for high-scoring teachers. There were also consequences for school administrators based on their component of the total value-added amount.

There are fundamentally two approaches taken in dealing with the inevitable missing data. One is to only deal with students who have complete data and draw inferences as if they were representative of all the students (missing at random). A more sophisticated approach that is used is to impute the missing values based on the scores of the students who had scores, perhaps matched on other information that was available. Inferences from either approach have limitations, sometimes requiring what Harvard's Don Rubin characterized as "heroic assumptions."

To make the problems of such a missing data strategy more vivid, I suggested (tongue firmly in cheek) that were I a principal in a school being evaluated I would take advantage of the imputation scheme by having an enriching, educational field trip for the top half of the students on the day of the pretest and another, parallel one, for the bottom half on the day of the posttest. The missing groups would have scores imputed

[1] http://investigations.nbcnews.com/_news/2013/03/20/17376664-caught-cheating-colleges- falsify-admissions-data-for-higher-rankings (accessed August 24, 2015).

for them based on the average of the scores of those who were there. Such a scheme would boost the change scores, and the amount of the increase would be greatest for schools with the most diverse populations. Surely a win-win situation.

Whenever I gave a talk about value-added and mentioned this scheme to game the school evaluations it always generated guffaws from most of the audience (although there were always a few who busied themselves taking careful notes). I usually appended the *obiter dictum* that if I could think of this scheme, the school administrators in the field, whose career advancement was riding on the results, would surely be even more inventive. Sadly, I was prescient.

On October 13, 2012, Manny Fernandez reported in *the New York Times* that Former El Paso schools superintendent, Lorenzo Garcia was sentenced to prison for his role in orchestrating a testing scandal. The Texas Assessment of Knowledge and Skills (TAKS) is a state-mandated test for high school sophomores. The TAKS missing data algorithm was to treat missing data as missing at random, and hence the score for the entire school was based solely on those who showed up. Such a methodology is so easy to game that it was clearly a disaster waiting to happen. And it did. The missing data algorithm used by Texas was obviously understood by school administrators; for all aspects of their scheme was to keep potentially low-scoring students out of the classroom so they would not take the test and possibly drag scores down. Students identified as likely low performing "were transferred to charter schools, discouraged from enrolling in school or were visited at home by truant officers and told not to go to school on test day."

But it didn't stop there. Some students had credits deleted from transcripts or grades changed from passing to failing so they could be reclassified as freshman and therefore avoid testing. Sometimes students who were intentionally held back were allowed to catch up before graduation with "turbo-mesters" in which a student could acquire the necessary credits for graduation in a few hours in front of a computer.

Superintendent Garcia boasted of his special success at Bowie High School, calling his program "the Bowie Model." The school and its administrators earned praise and bonuses in 2008 for its high rating. Parents and students called the model "*los desaparecidos*" (the disappeared). It received this name because in the fall of 2007 381 students were enrolled in Bowie as freshman; however, the following fall the sophomore class was composed of only 170 students.[2]

It is an ill wind indeed that doesn't blow some good. These two examples contain the germ of good news. While the cheating methodologies employed utilize the shortcomings of the missing data schemes that were in use to game the system, they also tell us two important things:

(1) Dealing with missing data is a crucial part of any practical situation, and doing it poorly is not likely to end well; and
(2) Missing data methodologies are not so arcane and difficult that the lay public cannot understand them.

So we should not hesitate to employ the sorts of full-blooded methods of multiple imputations pioneered by Rod Little and Don Rubin; opponents cannot claim that they are too complicated for ordinary people to understand. The unfolding of events has shown conclusively their general comprehensibility. But this is not likely to be enough. To be effective we almost surely need to use some sort of serious punishment for those who are caught.[3]

[2] http://www.elpasotimes.com/episd/ci_20848628/former-episd-superintendent-lorenzo-garcia-enter-plea-aggreement (accessed August 24, 2015).

[3] This conclusion was sent to me in an e-mail from Don Rubin on December 12, 2013. He wrote, "Amazing how simple-minded these manipulators are. The only way to address these games is to use an IRS-like hammer = big punishment if you get caught."

SECTION II

# Communicating Like a Data Scientist

## Introduction

Princeton's polymath John Tukey (1915–2000) often declared that a graph is the best, and sometimes the only, way to find what you were not looking for. Tukey was giving voice to what all data scientists now accept as gospel – statistical graphs are powerful tools for the discovery of quantitative phenomena, for their communication, and even for the efficient storage of information.

Yet, despite their ubiquitousness in modern life, graphical display is a relatively modern invention. Its origins are not shrouded in history like the invention of the wheel or of fire. There was no reason for the invention of a method to visually display data until the use of data, as evidence, was an accepted part of scientific epistemology. Thus it isn't surprising that graphs only began to appear during the eighteenth-century Enlightenment after the writings of the British empiricists John Locke (1632–1704), George Berkeley (1685–1753), and David Hume (1711–76) popularized and justified empiricism as a way of knowing things.

Graphical display did not emerge from the preempirical murk in bits and pieces. Once the epistemological ground was prepared, its birth was more like Botticelli's Venus – arising fully adult. The year 1786 is the birthday of modern statistical graphics, devised by the Scottish iconoclast William Playfair (1759–1823) who invented[1] what was an almost entirely

[1] *Invented* may be a trifle too strong because there were scattered examples prepared previously, most notably in the study of weather, but they tended to be ad hoc primitive affairs, whereas Playfair's depictions were fully developed and, even by modern standards, beautiful and well designed.

new way to communicate quantitative phenomena. Playfair showed the rise and fall over time of imports and exports between nations with line charts; the extent of Turkey that lay in each of three continents with the first pie chart; and the characteristics of Scotland's trade in a single year with a bar chart. Thus in a single remarkable volume, an atlas that contained not a single map, he provided spectacular versions of three of the four most important graphical forms. His work was celebrated and has subsequently been seized upon by data scientists as a crucial tool to communicate empirical findings to one another and, indeed, even to oneself.

In this section we celebrate graphical display as a tool to communicate quantitative evidence by telling four stories. In Chapter 8 we set the stage with a discussion of the empathetic mind-set required to design effective communications of any sort, although there is a modest tilting toward visual communications. The primary example shows how communicating sensitive material, in this case the result of genetic testing for mutated genes that increase the likelihood of cancer, requires both empathy and an understanding of basic graphic design principles.

In Chapter 9 we examine the influence that graphs designed by scientists have influenced the media, and vice versa, commenting specifically on how media designers seem, at least at this moment in history, to be doing a better job – but not always.

The history of data display has provided innumerable examples of clear depictions of two-dimensional data arrayed over a two-dimensional surface (a map depicting the locally two-dimensional surface of the Earth is the earliest and best example). The design challenge is how to depict more than two dimensions on that same flat surface. Shading a map to depict population is a popular approach to show a third dimension on top of the two geographic ones. But what about four dimensions? Or five? Or more? Evocative solutions to this challenge are justly celebrated (the most celebrated is Charles Joseph Minard's 1869 six-dimensional depiction of Napoleon's catastrophic Russian campaign). In Chapter 10 we introduce the inside-out plot as a way of exploring very high-dimensional data, and then, in Chapter 11 we drop back almost two centuries and introduce the nineteenth-century English reformer Joseph Fletcher, who plotted moral statistics on top of the two geographic variables to illustrate the distribution of variables like crime, ignorance, bastardy, and improvident marriages across England. By juxtaposing these "moral maps" he tried to elicit causal conclusions (e.g., areas of high ignorance were also areas of high crime) and so suggest how alleviating one might have a positive effect on the other (e.g., increased funding of education would reduce crime). After introducing some of Fletcher's maps derived from 1835 data we show how well the suggested reforms worked by using the same methods but with more recent data. We also transform Fletcher's moral maps into a more modern graphic form and show the qualitative arguments that Fletcher was making become much more vivid and that the strength of the relationship between the variables could be estimated quantitatively using scatter plots.

# 8

# On the Crucial Role of Empathy in the Design of Communications

## *Genetic Testing as an Example*

*Good information design is clear thinking made visible, while bad design is stupidity in action.*

Edward Tufte 2000

*An effective memo should have, at most, one point.*

Paul Holland 1980

The effectiveness of any sort of communication depends strongly on the extent to which the person preparing the communication is able to empathize with its recipients. To maximize effectiveness it is absolutely crucial to determine what information the receiver needs to hear and not let that be overwhelmed by other things that you may want to tell her.

We get a sense of how hard it is to do this when we see how celebrated are those rare communications that do it well. More than a dozen years ago my son received a letter from Princeton University in response to his application. That letter, reproduced in the following text, exemplifies the empathetic attitude I recommend. Fred Hargadon, then Princeton's dean of admission, recognized that the recipient is principally interested in just one thing, and he designed his famous letter to focus on just that (see Figure 8.1).

PRINCETON UNIVERSITY Admission Office
MAILING ADDRESS: Box 430, Princeton, New Jersey 08544-0430
OFFICE: 110 West College
TELEPHONE: 609-258-3060 FACSIMILE: 609-258-6743

Fred A. Hargadon
Dean of Admission
(on leave until January 2001)

Steve LeMenager
Acting Dean of Admission

December, 2000

**Dear Sam,**

# Yes!

**We are happy to offer you admission to Princeton University and are delighted to welcome you as a member of the class of 2005.**

**Sincerely,**

**Steve LeMenager**
**Acting Dean of Admissions**

FIGURE 8.1. Acceptance letter from Princeton University.

Only 8 percent of Princeton's more than twenty thousand applicants received such a letter, but it isn't hard to imagine the character of a parallel version (imagined in Figure 8.2) that might have been sent to the remaining 92 percent.

Of course, Princeton did not send such a letter. I was told that the parallel version I ginned up was never in the cards. When I asked about the character of the actual "No!" letter, Dean LeMenager told me "You can say 'Yes' quickly; 'No' takes a lot longer."

Letters announcing the outcome of an application to a selective university or offering a job are but two of the many important situations in which information is conveyed to a candidate. Crafting such messages is rarely done as thoughtfully as Princeton's remarkable letter. But these circumstances, as important as they are, pale in comparison with other

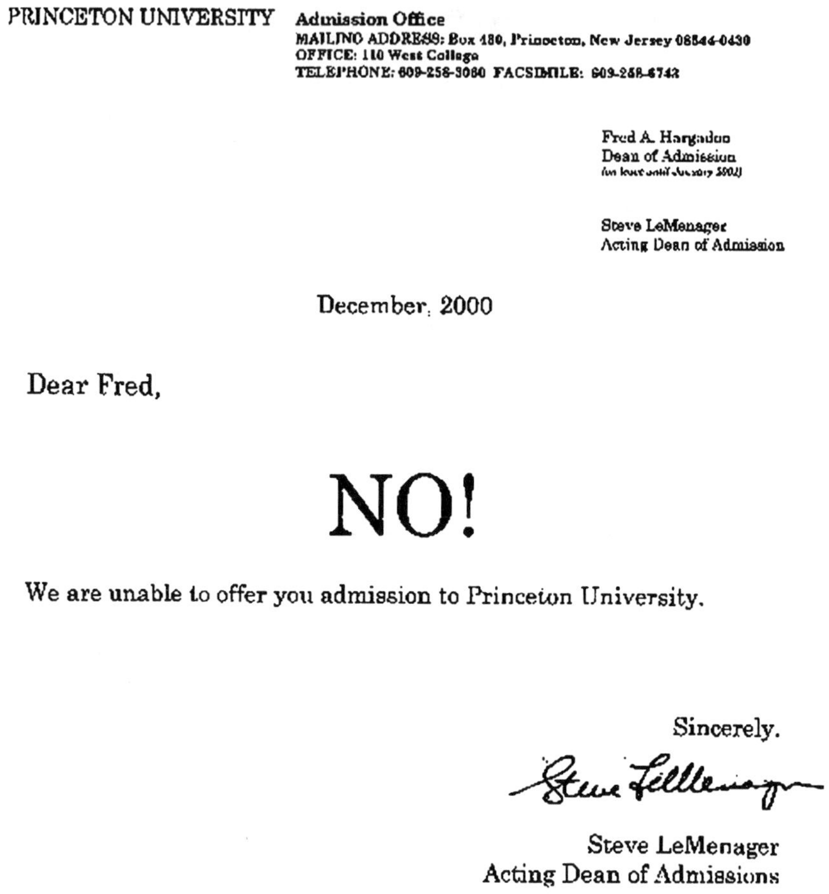

PRINCETON UNIVERSITY **Admission Office**
MAILING ADDRESS: Box 430, Princeton, New Jersey 08544-0430
OFFICE: 110 West College
TELEPHONE: 609-258-3060 FACSIMILE: 609-258-6743

Fred A. Hargadon
Dean of Admission
(on leave until January 2001)

Steve LeMenager
Acting Dean of Admission

December, 2000

Dear Fred,

# NO!

We are unable to offer you admission to Princeton University.

Sincerely,

Steve LeMenager
Acting Dean of Admissions

FIGURE 8.2. Faux rejection letter from Princeton University that parallels the form of Figure 8.1.

kinds of communications. One of these arose on May 14, 2013, when Angelina Jolie wrote an OpEd essay in the *New York Times* telling of her decision to have a double mastectomy.

The trail that ended with this drastic decision began many years earlier, when her forty-six-year-old mother contracted breast cancer, which she succumbed to ten years later. Ms. Jolie feared a familial component and decided to be tested and find out whether she carried a genetic mutation that would substantially increase the likelihood of her suffering the same fate as her mother.

Ordinarily, one in eight women will contract breast cancer during their lives, but this 12 percent chance is increased six- to sevenfold if a woman carries a mutated form of the 187delAG BRCA1, 5385insC BRCA1, or 617delT BRCA2 genes. Such mutations are, thankfully, very rare, but among Ashkenazi Jewish women the likelihood of such a mutation rises to about 2.6 to 2.8 percent. In addition, there are other risk factors that also raise the likelihood of such a mutation.

Ms. Jolie had the genetic test and discovered that she carried the unfortunate mutation, and, on that basis, decided to have a prophylactic double mastectomy.

How do you communicate the results of such a test? If ever a situation required empathy, this is it. This situation differs from admission to Princeton in that more than 97 percent of the time the message carries happy news. It is the remaining small part of the population that requires special handling. But let us start with the report that carries good news (see Figure 8.3).

Although it contains a great deal of verbiage, the report resembles the Princeton letter in at least one aspect: the largest type size is reserved for the summary message of principal interest to the recipient. I suspect that this report is the product of many committees made up of geneticists, physicians, counselors, and lawyers who met over a substantial period of time. Inevitably, such collaborations generate a kind of entropy that keeps adding more while subtracting nothing.

We can learn from Princeton's letter and emphasize even more what the recipient wants to know. One display that responds to this enlarges the basic positive message and moves the details to the back to be perused later, should the recipient ever care to do so (see Figure 8.4).

But what about the other letter? It goes to fewer than 3 percent of the people tested, but the news it carries portends a nightmare. Dean LeMenager's wise advice looms large. The good news we can say fast; the bad news ought to take longer. What does this letter currently look like (see Figure 8.5)?

Surprisingly, in form it is identical to the one that carries the happy news of no mutations. Only the content varies. Is this the best we can do?

**CONFIDENTIAL**

***Integrated BRACAnalysis®***
***BRCA1* and *BRCA2* Analysis Result**

| PHYSICIAN | SPECIMEN | | PATIENT | |
|---|---|---|---|---|
| **John Smith, MD** | Specimen: | **Blood** | Name: | **Doe, Jane** |
| **Comprehensive Medical Center** | Draw date: | **Aug 01, 2010** | Date of Birth: | **April 1, 1492** |
| **1100 Grand Ave** | Accession date: | **Aug 02, 2010** | Patient ID: | **000000** |
| **Away, GA 12345** | Report Date: | **Jun 22, 2011** | Gender: | **Female** |
| | | | Accession #: | **00000000-BLD** |
| | | | Requisition #: | **000000** |

**Test Results and Interpretation**

**NO MUTATION DETECTED**

| Test Performed: | Result: | Interpretation: |
|---|---|---|
| *BRCA1* sequencing | No Mutation Detected | No Mutation Detected |
| comprehensive rearrangement | No Mutation Detected | No Mutation Detected |
| *BRCA2* sequencing | No Mutation Detected | No Mutation Detected |
| comprehensive rearrangement | No Mutation Detected | No Mutation Detected |

It is our understanding that this patient was identified for testing due to a personal or family history suggestive of hereditary breast and ovarian cancer. Analysis consists of sequencing of all translated exons and immediately adjacent intronic regions of the BRCA1 and BRCA2 genes and a comprehensive rearrangement test of both BRCA1 and BRCA2 by quantitative PCR analysis (BRACAnalysis Rearrangement Test, BART). The classification and interpretation of all variants identified in this assay reflects the current state of scientific understanding at the time this report was issued. In some instances, the classification and interpretation of such variants may change as new scientific information becomes available.

No deleterious mutation was found in BRCA1 or BRCA2 in this individual by sequencing and quantitative PCR analysis. This test is designed to identify mutations in 22 exons and approximately 750 adjacent intronic base pairs of BRCA1 as well as 26 exons and approximately 950 adjacent intronic base pairs of BRCA2 (a total of over 17,600 base pairs analyzed). This test is also designed to detect duplications and deletions involving the promoter region and coding exons of BRCA1 and BRCA2. There are other, rare genetic abnormalities in BRCA1 and BRCA2 that this test will not detect. This result, however, rules out the majority of abnormalities believed to be responsible for hereditary susceptibility to breast and ovarian cancer (Ford D et al., Am J Human Genetics 62:676-689, 1998). If this individual has never had breast or ovarian cancer, it is recommended that testing an affected relative be considered to help clarify the clinical significance of this individual's negative test result.



Please contact Myriad Professional Support at 1-800-469-7423 to discuss any questions regarding this result.

Director Name Here
Qualifications Here

Director Name Here
Qualifications Here

These test results should only be used in conjunction with the patient's clinical history and any previous analysis of appropriate family members. It is strongly recommended that these test results be communicated to the patient in a setting that includes appropriate counseling. The accompanying Technical Specifications summary describes the analysis, method, performance characteristics, nomenclature, and interpretive criteria of this test. This test may be considered investigational by some states. This test and its performance characteristics were determined by Myriad Genetic Laboratories. It has not been reviewed by the U.S. Food and Drug Administration. The FDA has determined that such clearance or approval is not necessary.

FIGURE 8.3. Notice of negative finding of gene mutation from Myriad Laboratories.

Before considering changes, if any, it is crucial that we place this report in its proper context. The patient does not receive the report in a vacuum. It is delivered by a genetic counselor, who explains what everything means and what are the various options available. Usually

**CONFIDENTIAL**

***Integrated BRACAnalysis®***
***BRCA1* and *BRCA2* Analysis Result**

| PHYSICIAN | SPECIMEN | | PATIENT | |
|---|---|---|---|---|
| **John Smith, MD** | Specimen: | **Blood** | Name: | **Doe, Jane** |
| **Comprehensive Medical Center** | Draw date: | **Aug 01, 2010** | Date of Birth: | **April 1, 1492** |
| **1100 Grand Ave** | Accession date: | **Aug 02, 2010** | Patient ID: | **000000** |
| **Away, GA 12345** | Report Date: | **Jun 22, 2011** | Gender: | **Female** |
| | | | Accession #: | **00000000-BLD** |
| | | | Requisition #: | **000000** |

**Test Results and Interpretation**

**NO MUTATION DETECTED**

| Test Performed: | Result: | Interpretation: |
|---|---|---|
| *BRCA1* sequencing | No Mutation Detected | No Mutation Detected |
| comprehensive rearrangement | No Mutation Detected | No Mutation Detected |
| *BRCA2* sequencing | No Mutation Detected | No Mutation Detected |
| comprehensive rearrangement | No Mutation Detected | No Mutation Detected |

FIGURE 8.4. Suggested revision of notice of negative finding of gene mutation that emphasizes the principal message.

an oncologist, who can discuss the medical options in detail, later joins the counselor. In the grand scheme of things, such extra help is neither practical nor necessary for Princeton applicants.

Remembering that the core of effective communication is empathy, let us try to imagine what it is like for patients who have undergone genetic testing and are awaiting the results. After the test they are given an appointment a fortnight or so in the future and asked to come in then for the results. At the appointed hour they arrive and sit nervously in a large waiting room. Often a loved one accompanies them, as they await, in terror, what might be the potential outcome. After what must seem like an eternity they are led to an examining room, where they wait until a serious-faced genetic counselor comes in with a folder. She sits down, opens the folder, and delivers the news.

The vast majority of the time, the test reveals no mutation, and although the counselor is careful to point out that sometimes, albeit rarely, the test has missed a mutation, the mood is celebratory, and the associated caveats fade into the background. After a short time, perhaps only after leaving the clinic, the thought occurs: "Why did I have to come in? Why couldn't someone call as soon as the results were available and tell me 'everything's OK'?"

**CONFIDENTIAL**

**Comprehensive BRACAnalysis®**
***BRCA1* and *BRCA2* Analysis Result**

| PHYSICIAN | SPECIMEN | | PATIENT | |
|---|---|---|---|---|
| **John Smith, MD** | Specimen: | **Blood** | Name: | **Doe, Jane** |
| **Comprehensive Medical Center** | Draw date: | **Aug 01, 2010** | Date of Birth: | **April 1, 1492** |
| **1100 Grand Ave** | Accession date: | **Aug 02, 2010** | Patient ID: | **000000** |
| **Away, GA 12345** | Report Date: | **Jun 22, 2011** | Gender: | **Female** |
| | | | Accession #: | **00000000-BLD** |
| | | | Requisition #: | **000000** |

**Test Results and Interpretation**

**POSITIVE FOR A DELETERIOUS MUTATION**

| Test Performed: | Result: | Interpretation: |
|---|---|---|
| *BRCA1* sequencing | No Mutation Detected | No Mutation Detected |
| 5-site rearrangement panel | No Mutation Detected | No Mutation Detected |
| *BRCA2* sequencing | S1970X (6137C>A) | **Deleterious** |

It is our understanding that this patient was identified for testing due to a personal or family history suggestive of hereditary breast and ovarian cancer. Analysis consists of sequencing of all translated exons and immediately adjacent intronic regions of the BRCA1 and BRCA2 genes and a test for five specific BRCA1 rearrangements. There are additional large genomic rearrangements in BRCA1 and in BRCA2, which are not detected by this test, but can be identified with the BRACAnalysis Rearrangement Test (BART). The classification and interpretation of all variants identified in this assay reflects the current state of scientific understanding at the time this report was issued. In some instances, the classification and interpretation of such variants may change as new scientific information becomes available.

The results of this analysis are consistent with the germline BRCA2 mutation S1970X, resulting in premature truncation of the BRCA2 protein at amino acid position 1970. Although the exact risk of breast and ovarian cancer conferred by this specific mutation has not been determined, studies of this type of mutation in high-risk families indicate that deleterious mutations in BRCA2 may confer as much as an 84% risk of breast cancer and a 27% risk of ovarian cancer by age 70 in women (Am. J. Hum. Genet. 62:676-689, 1998). Mutations in BRCA2 have been reported to confer a 12% risk of a second breast cancer within five years of the first (J Clin Oncol 17:3396-3402, 1999), as well as a 16% risk of subsequent ovarian cancer (J Natl Cancer Inst 91:1310-1315, 1999). Additionally, studies have shown that BRCA2 mutations confer as much as a 7% risk of pancreatic cancer by age 80 (J Med Genet 42:711-9, 2005); however, this risk may be higher in families in which pancreatic cancer has previously been diagnosed (Cancer Res 64:2634-2638, 2004). This mutation may also confer up to an 8% risk of male breast cancer and 20% risk of prostate cancer by age 80 (J Natl Cancer Inst 99:1811-4, 2007; J Natl Cancer Inst 91:1310-1315, 1999), as well as increased (albeit low) risks of some other cancers. Each first degree relative of this individual has a one-in-two chance of having this mutation. Family members can be tested for this specific mutation with a single site analysis.

Please contact Myriad Professional Support at 1-800-469-7423 to discuss any questions regarding this result.

Director Name Here
Qualifications Here

Director Name Here
Qualifications Here

These test results should only be used in conjunction with the patient's clinical history and any previous analysis of appropriate family members. It is strongly recommended that these test results be communicated to the patient in a setting that includes appropriate counseling. The accompanying Technical Specifications summary describes the analysis, method, performance characteristics, nomenclature, and interpretive criteria of this test. This test may be considered investigational by some states. This test and its performance characteristics were determined by Myriad Genetic Laboratories. It has not been reviewed by the U.S. Food and Drug Administration. The FDA has determined that such clearance or approval is not necessary.



FIGURE 8.5. Notice of positive finding of gene mutation from Myriad Laboratories.

Why indeed? Let us now revert to the 3 percent (or fewer) who do not get good news. For them the world just got a great deal more complicated and more dangerous. The genetic counseling is more important; the oncological discussions are, quite literally, a matter of life and death.

Considering both groups, we can see that if most people were to get an early communication that says "All Clear," those who hear nothing, and hence must come into the clinic for their appointment, can infer the bad news, and hence know it in advance of having the support of a counselor to explain things.

The issue is clear. Is the reduced angst yielded by providing early news to the 97 percent who do not have a mutation overbalanced by the lack of immediate counseling for those with a positive report? This is not an easy calculation. Perhaps it is helpful to remember that when the news is bad, there are only two kinds of reports. One kind causes the recipient great sadness and terror, and the other does the same thing a little worse. One has no good way to convey this sort of information; one has only a bad way and a worse way. I am reminded of what the effect must have been during World War II on a family that received a telegram from the War Department. Even without opening it, they knew that their worst nightmare had been realized. How much difference would it have made had the telegram been replaced by a skilled clinical psychologist who walked up the front steps to deliver the news in person?

It seems a cogent argument could be made for a policy that schedules all patients for an appointment when results become available, but for those with a negative outcome a phone call could deliver the happy news and cancel the appointment. The balance would come in as scheduled. It surely merits further discussion.

I believe that a report akin to the alternative I sketched as Figure 8.4, which shouts the good news and leaves the technical details in a very subsidiary role, could be of value. It seems worthwhile to try such a change in format, as well as changing the time and mode of delivery. It is also abundantly clear that, for those who receive the news of a relevant mutant gene, no format change would make any material difference.

The story told in this chapter provides a dramatic example of the thought processes that ought to accompany all communications. Data are always gathered within a context, and to understand and communicate those data without seriously considering that context is to court disaster. For this reason it is always a mistake to "staff out" the preparation of data displays to an underling whose understanding of both the context of the data and its potential uses is incomplete. The smallest error that might emerge from such a practice is that you could miss what you might have found; more serious errors can easily be imagined, for example, consider the consequences if Princeton had actually sent out my poorly conceived rejection letter shown as Figure 8.2.

# 9

# Improving Data Displays

## *The Media's and Ours*

### Introduction

In the transactions between scientists and the media, influence flows in both directions.[1] More than thirty years ago[2] I wrote an article with the ironic title "How to Display Data Badly." In it I chose a dozen or so examples of flawed displays and suggested some paths toward improvement. Two major newspapers, the *New York Times* and the *Washington Post*, were the source of most of my examples. Those examples were drawn over a remarkably short period of time. It wasn't hard to find examples of bad graphs.

Happily, in the intervening years those papers have become increasingly aware of the canons of good practice and have improved their data displays profoundly. Indeed, when one considers both the complexity of the data that are often displayed as well as the short time intervals permitted for their preparation, the results are often remarkable.

Eight years ago, over the course of about a fortnight, I picked out a few graphs from the *New York Times* that were especially notable. Over

[1] This is the third incarnation of this essay. In 2007 it appeared as an article in *CHANCE*, a magazine aimed at statisticians; a revised version was published as chapter 11 in my 2009 book *Picturing the Uncertain World*. In 2015, as I was preparing this book, I decided that the essay's message was even more relevant today than it was when it was originally prepared and so opted to include the revised version you see here. I am grateful to *CHANCE* and Princeton University Press for their permission to reuse this material.

[2] Wainer 1984.

the same period I noticed decidedly inferior graphs in the scientific literature for data that had the same features. At first I thought that it felt more comfortable in the "good old days" when we scientists did it right and the media's results were flawed. But the old days were not actually so good. Graphical practices in scientific journals have not evolved as fast as those of the mass media. This year I redid the same investigation and reached the same conclusions. It is time we in the scientific community learned from the media's example.

## Example 1: Pies

The U.S. federal government is fond of producing pie charts, and so it came as no surprise to see (Figure 9.1) a pie chart of the sources of government receipts. Of course, the grapher felt it necessary to "enliven" the presentation by adding a specious extra dimension, and to pull forward

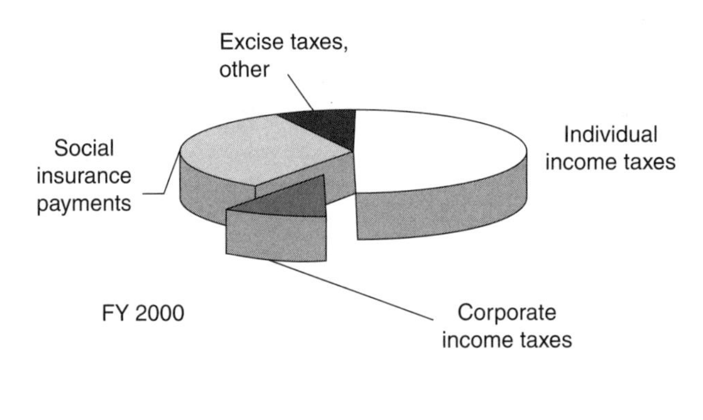


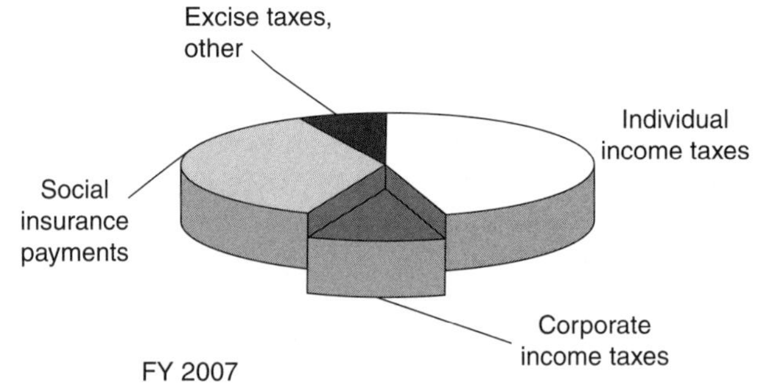


FIGURE 9.1. A typical 3D pie chart. This one, no worse nor better than most of its ilk, is constructed from U.S. government budget data (http://www.whitehouse.gov/omb/budget/fy2008/pdf/hist.pdf; accessed December 18, 2008).

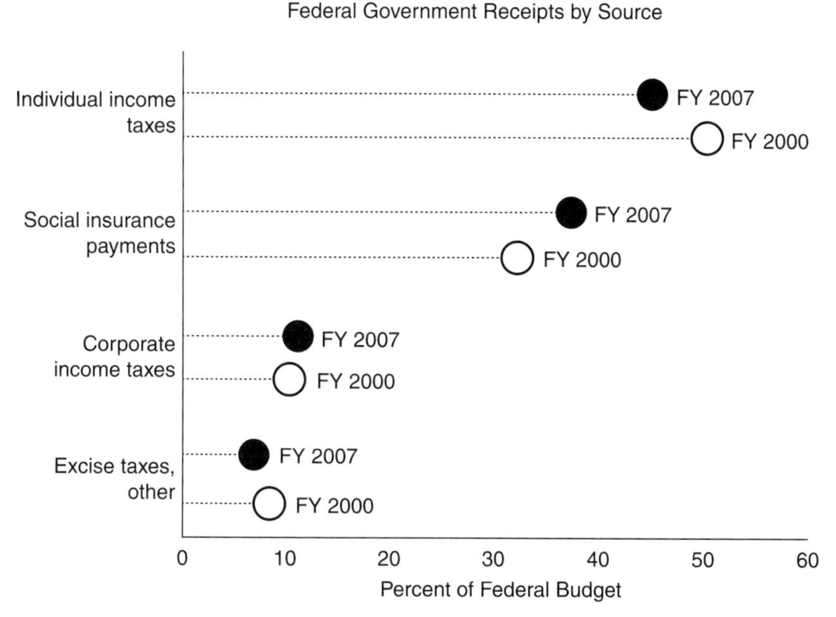


FIGURE 9.2. A restructuring of the same data shown in Figure 9.1 making clearer what changed in the sources of government receipts between the year 2000 and 2007.

the segment representing corporate taxes, which has the unfortunate perceptual consequence of making that segment look larger than it is. Presenting the results for 2000 and 2007 together must have been to allow the viewer to see the changes that have taken place over that time period (roughly the span of the Bush administration). The only change I was able to discern was shrinkage in the contribution of individual income taxes.

I replotted the data in a format that provides a clearer view (Figure 9.2) and immediately saw that the decrease in individual income taxes was offset by an increase in social insurance payments. More specifically, increasing social security taxes, whose effect ends after the first hundred thousand dollars of earned income, paid for the cost of tax cuts aimed principally at the wealthy.

I wasn't surprised that such details were obscured in the original display, for my expectation of data displays constructed for broad consumption was not high. Hence when I was told of a graph in an article in the *New York Times Sunday Magazine* about the topics that American clergy choose to speak out on, I anticipated the worst. My imagination created,

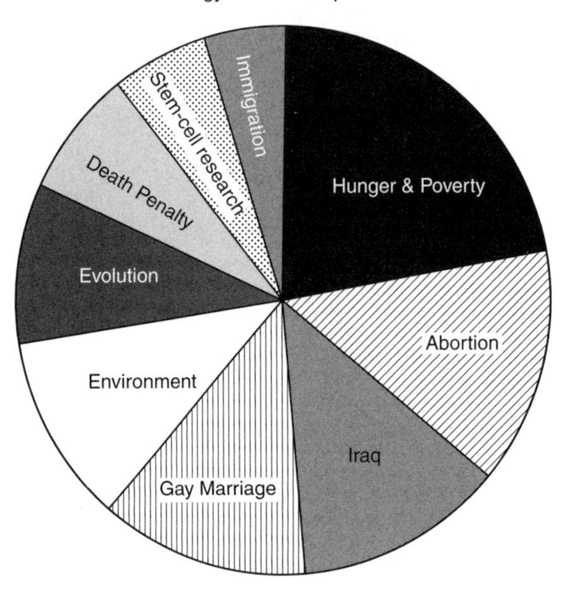


FIGURE 9.3. A typical pie chart representation of the relative popularity of various topics among the U.S. clergy.

floating before my eyes, a pie chart displaying such data (Figure 9.3). My imagination can only stretch so far, hence the pie I saw had only two dimensions, and the categories were ordered by size.

Instead, I found a pleasant surprise (Figure 9.4).[3] The graph (produced by an organization named "Catalogtree") was a variant of a pie chart in which the segments all subtend the same central angle, but their radii are proportional to the amount being displayed. This is a striking improvement on the usual pie because it facilitates comparisons among a set of such displays representing, for example, different years or different places. The results from each segment are always in the same place, whereas with pie charts the locations of segments may vary as the data change. Compare it with the pie in Figure 9.1, in which one segment leapt out for no apparent reason, except possibly to mislead. In Figure 9.4 the extended segment that represents the topics of hunger and poverty is

[3] Rosen 2007.

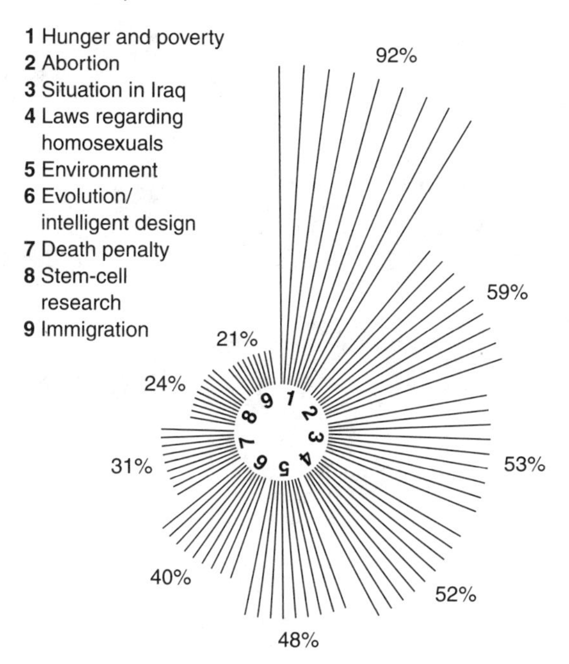


FIGURE 9.4. A display from the February 18, 2007 *New York Times Sunday Magazine* (page 11) showing the same data depicted in Figure 9.3 as a Nightingale Rose.

eye-catching for a good reason – it represents the single topic that dominates all others discussed in church.

This plot also indicates enough historical consciousness to evoke Florence Nightingale's (1858) famous Rose of the Crimean War (Figure 9.5).[4] The original Nightingale Rose dramatically showed the far greater death toll caused by lack of sanitation than battle wounds and was

[4] Also chapter 11 in Wainer 2000.

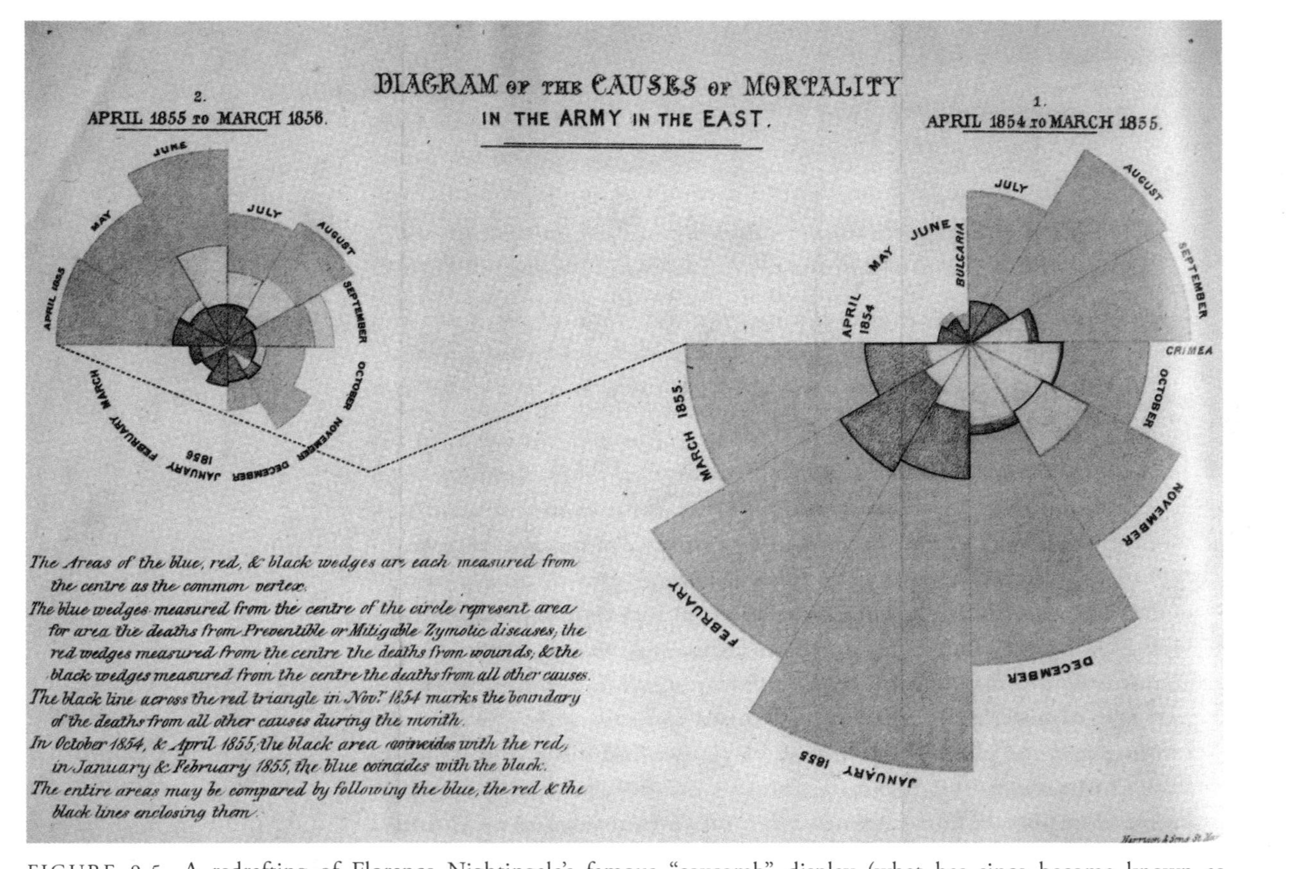


FIGURE 9.5. A redrafting of Florence Nightingale's famous "coxcomb" display (what has since become known as a Nightingale Rose) showing the variation in mortality over the months of the year.

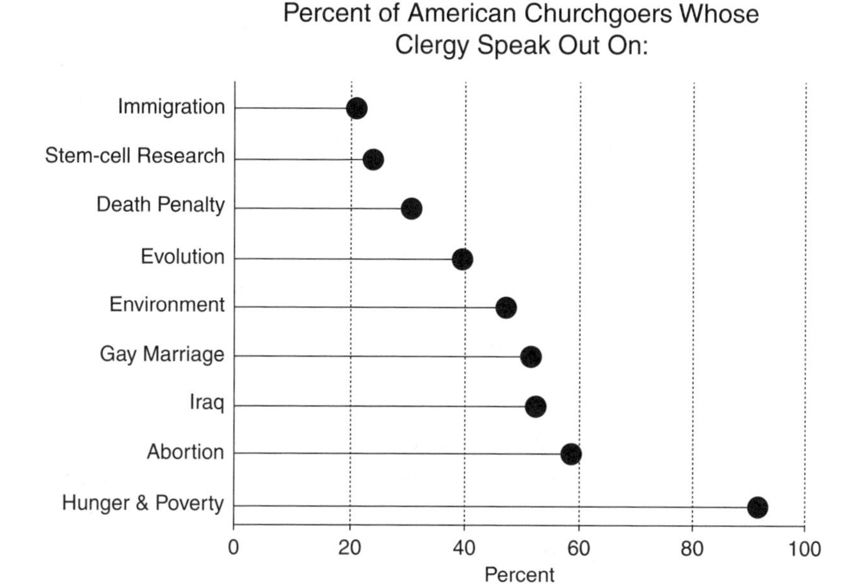


FIGURE 9.6. The same data previously shown in Figures 8.3 and 8.4 recast as a line-and-dot plot.

hence very effective in helping her reform the military ministry's battlefield sanitation policies.

Sadly, this elegant display contains one small flaw that distorts our perceptions. The length of the radius of each segment is proportional to the percentage depicted; but the area of the segment, not its radius, influences us. Thus, the radii need to be proportional to the square root of the percentage for the areas to be perceived correctly. An alternative figuration with this characteristic is shown as Figure 9.6.

Journalists might complain that the staid nature of Figure 9.6 does not make the visual point about hunger and poverty being the big story as emphatically as does the *New York Times* original (Figure 9.4). No it doesn't. And that is exactly my point. Although hunger and poverty are, in combination, the most common topic, it does not dominate the way our perceptions of Figure 9.4 suggest. The goal of good display is full and accurate communication and so a display that yields misperceptions is a graphical lie.

## Example 2: Line Labels

In 1973 Jacques Bertin, the acknowledged master theorist of modern graphics explained that when one produces a graph, it is best to label each of the elements in the graph directly. He proposed this as the preferred alternative to appending some sort of legend that defines each element. His point was that when the two are connected, you could comprehend the graph in a single moment of perception, as opposed to having to first look at the lines, then read the legend, and then match the legend to the lines.

This advice is too rarely followed. For example, Michigan State's Mark Reckase,[5] in a simple plot of two lines (Figure 9.7), chose not to label the lines directly – even though there was plenty of room to do so – and instead chose to put in a legend. And the legend reverses the order of the lines, so the top line in the graph becomes the bottom line in the legend, thus increasing the opportunity for reader error.

Can the practice of labeling be made still worse? Figure 9.8, from Pfeffermann and Tiller,[6] comes a valiant effort to do so. Here the legend is hidden in the figure caption, and again its order does not match the order of the lines in the graph. Moreover, the only way to distinguish BMK from UnBMK is to notice a small dot. The only way I could think of that would make the connection between the various graphical elements and their identifiers worse would be to move the latter to an appendix.

How does the *New York Times* fare on this aspect of effective display? Very well indeed. In Figure 9.9 are two plots roughly following a *New York Times* design that describe one hundred years of employment in New Haven County. In each panel the lines are labeled directly, making the decline of manufacturing jobs clear. In the following week another graph appeared showing five time series over three decades. A redrafted and corrected version of a *New York Times* design is in Figure 9.10. In this plot the long lines and their crossing patterns made it possible for the viewer to confuse one line with another. Labeling both ends of each line ameliorated this possibility; a fine idea, worthy of being copied by those of us whose data share the same characteristics. Unfortunately, in

[5] Reckase 2006.
[6] Pfeffermann and Tiller 2006.

FIGURE 9.7. A graph taken from Reckase (2006) in which instead of identifying the graph lines directly they are identified through a legend – indeed a legend whose order does not match the data.

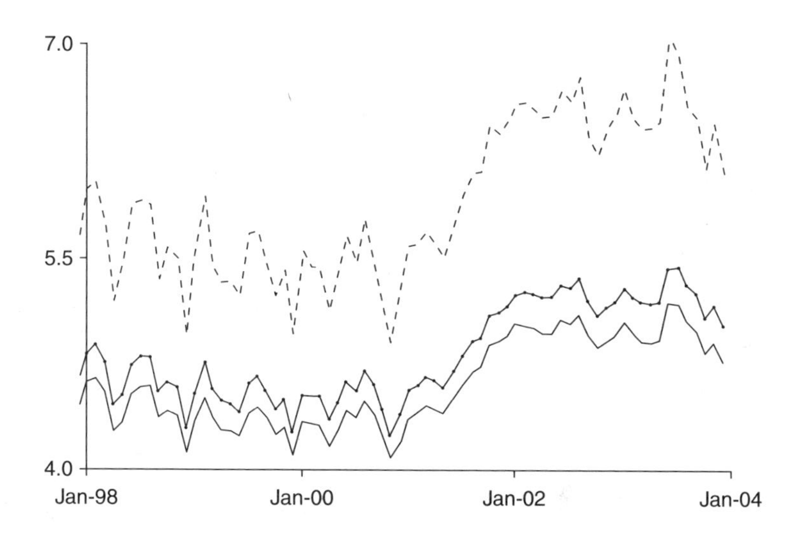


*STD of CPS, Benchmarked, and Unbenchmarked Estimates of Total Monthly Unemployment, South Atlantic Division (numbers in 10,000) (- - - - CPS; —— BMK; ——•—— UnBMK).*

FIGURE 9.8. A graph taken from Pfeffermann and Tiller (2006) in which the three data series are identified in acronym form in the caption. There is plenty of room on the plot for them to be identified on the graph with their names written out in full.

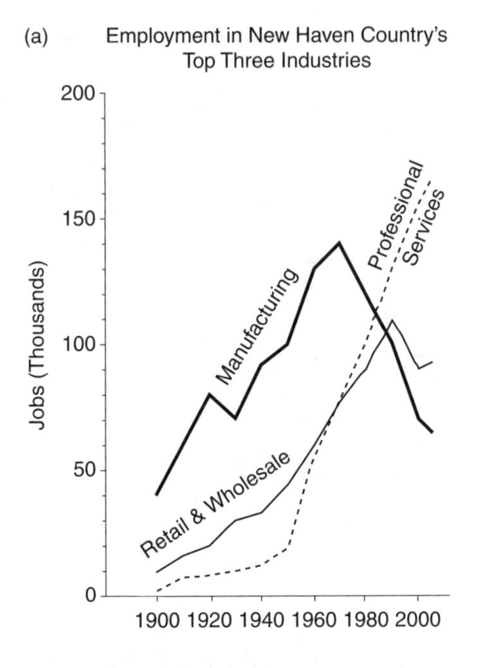


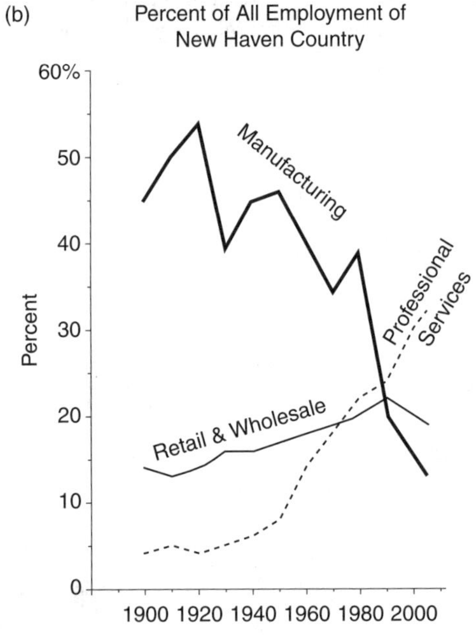


FIGURE 9.9. A graph modeled after one from the Metro section of the February 18, 2007 *New York Times* showing two panels containing three lines each, in which each line is identified directly. See panels 9.9a and 9.9b.

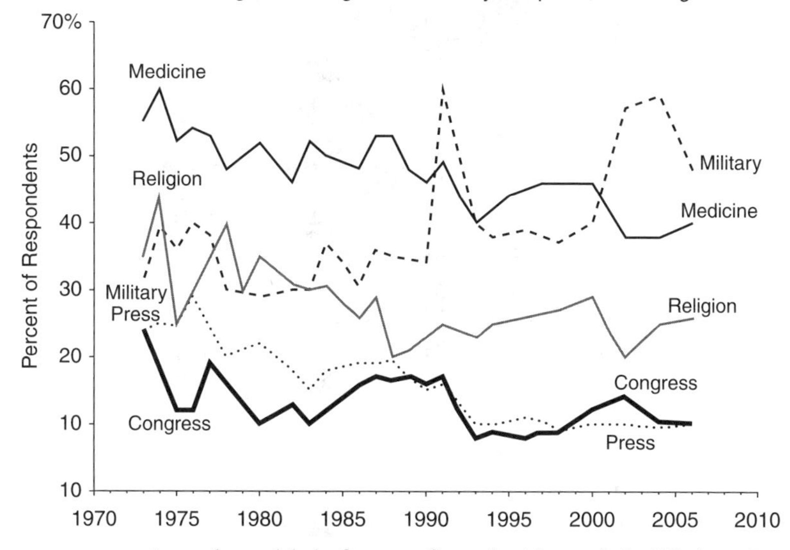


FIGURE 9.10. A graph modeled after one from the News of the Week in Review section of the February 25, 2007 *New York Times* (page 15) showing five long lines each, in which each line is identified directly at both of its extremes, thus making identification easy, even when the lines cross.

the original the grapher added some confusion by using equally spaced tick marks on the horizontal axis for unequal time periods (sometimes one year apart, sometimes two).

## Example 3: Channeling Playfair to Measure China's Industrial Expansion

On the business pages of the March 13, 2007 *New York Times* was a graph used to support the principal thesis of an article on how the growth of the Chinese economy has yielded, as a concomitant, an enormous expansion of acquisitions. That graph, shown here as Figure 9.11, has two data series. The first shows the amount of money spent by China on external acquisitions from 1990 through 2006. The second time series shows the number of such acquisitions.

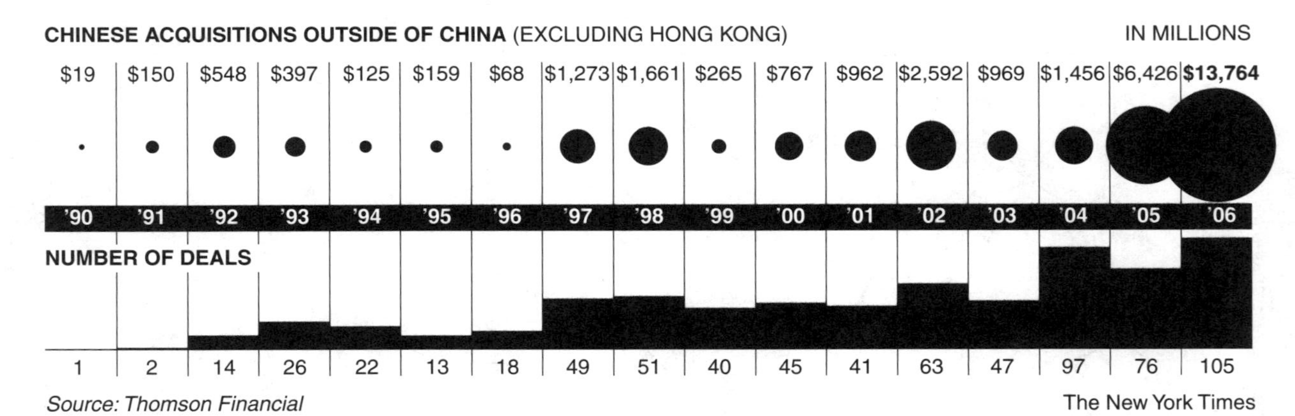


FIGURE 9.11. A graph taken from the business section of the March 13, 2007 *New York Times* (page C1) showing two data series. One is of a set of counts and is represented by bars; the second is money represented by the area of circles.

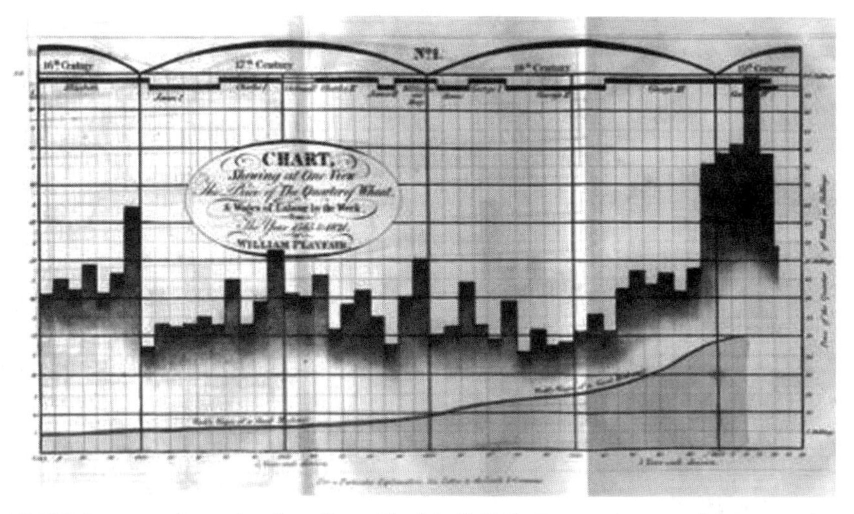


FIGURE 9.12. A graph taken from Playfair (1821). It contains two data series that are meant to be compared. The first is a line that represents the "weekly wages of a good mechanic" and the second is a set of bars that represent the "Price of a Quarter of Wheat."

The display format, while reasonably original in its totality, borrows heavily from William Playfair. First, the idea of including two quite different data series in the same chart is reminiscent of Playfair's 1821 chart comparing the cost of wheat with the salary of a mechanic (Figure 9.12).[7] However, in plotting China's expenditures the graphers had to confront the vast increases over the time period shown; a linear scale would have obscured the changes in the early years. The solution they chose was also borrowed from Playfair's plot of Hindoostan in his 1801 *Statistical Breviary* (Figure 9.13). Playfair showed the areas of various parts of Hindoostan as circles. The areas of the circles were proportional to the areas of the segments of the country, but the radii are proportional to the square root of the areas. Thus, by lining up the circles on a common line, we can see the differences of the heights of the circles that is, in effect, a square-root transformation of the areas. This visual transformation helps to place diverse data points on a more reasonable scale.

[7] Discussed in detail in Friendly and Wainer 2004.

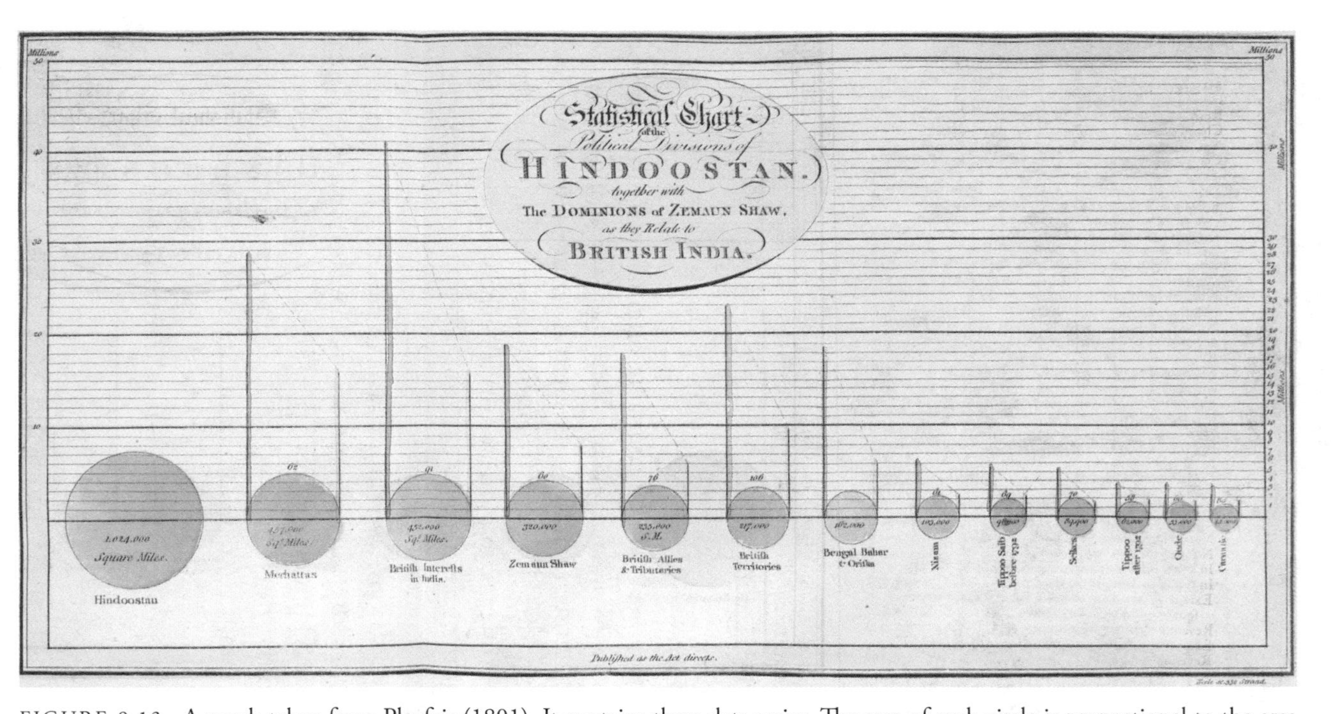


FIGURE 9.13. A graph taken from Playfair (1801). It contains three data series. The area of each circle is proportional to the area of the geographic location indicated. The vertical line to the left of each circle expresses the number of inhabitants, in millions. The vertical line to the right represents the revenue generated in that region in millions of pounds sterling.

The *New York Times*' s plot of China's increasing acquisitiveness has two things going for it. It contains thirty-four data points, which by mass media standards is data rich, showing vividly the concomitant increases in the two data series over a seventeen-year period. This is a long way from Playfair's penchant for showing a century or more, but in the modern world, where changes occur at a less leisurely pace than in the eighteenth century, seventeen years is often enough. And second, by using Playfair's circle representation it allows the visibility of expenditures over a wide scale.

Might other alternatives perform better? Perhaps. In Figure 9.14 is a two-paneled display in which each panel carries one of the data series. Panel 9.14a is a straightforward scatter plot showing the linear increases in the number of acquisitions that China has made over the past seventeen years. The slope of the fitted line tells us that over those seventeen years China has, on average, increased its acquisitions by 5.5/year. This crucial detail is missing from the sequence of bars but is obvious from the fitted regression line in the scatter plot. Panel 9.14b shows the increase in money spent on acquisitions over those same seventeen years. The plot is on a log scale, and its overall trend is well described by a straight line. That line has a slope of 0.12 in the log scale and hence translates to an increase of about 32 percent per year. Thus, the trend established over these seventeen years shows that China has both increased the number of assets acquired each year and also has acquired increasingly expensive assets.

The key advantage of using paired scatter plots with linearizing transformations and fitted straight lines is that they provide a quantitative measure of how China's acquisitiveness has changed. This distinguishes Figure 9.14 from the *New York Times* plot, which, although it contained all the quantitative information necessary to do these calculations, had primarily a qualitative message.

> Magnum esse solem philosophus probabit, quantus sit mathematicus.[8]
>
> *Seneca, Epistulae 88.27*

[8] Roughly translated, "while philosophy says the sun is large, mathematics takes its measure."

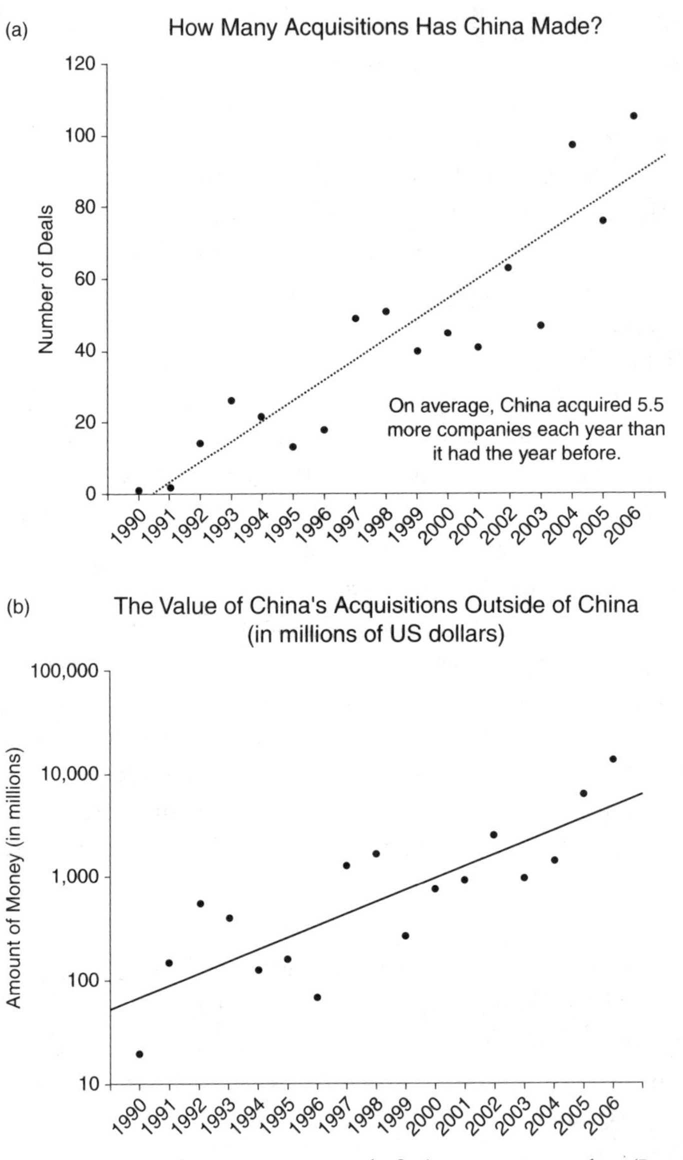


FIGURE 9.14. The data from Figure 9.11 redrafted as two scatter plots (Panel a and Panel b). The plot of money is shown on a log scale, which linearizes the relationship between the amounts and the passing years. The superposition of regression lines on both panels allows the viewer to draw quantitative inferences about the rates of growth that was not possible with the depiction shown in Figure 9.11.

## Conclusion

William Playfair set a high standard for effective data display more than two hundred years ago. Since that time rules have been codified,[9] and many books have described and exemplified good graphical practice.[10] All of these have had an effect on graphical practice. But it would appear from my sample of convenience, that the effect was larger on the mass media than on the scientific literature. I don't know why, but I will postulate two possible reasons. First, scientists make graphs with the software they have available and tend, more often than is proper, to accept the default options for that software. Producers of displays for large market mass media have, I suspect, a greater budget and more flexibility. The second reason why poor graphical practices persist is akin to Einstein's observation on the persistence of incorrect scientific theories: "Old theories never die, just the people who believe in them."[11]

Graphical display is prose's nonverbal partner in the quest to effectively communicate quantitative phenomena. When France's Louis XVI, an amateur of geography and owner of many fine atlases, first saw the statistical graphics invented by Playfair, he immediately grasped their meaning and their importance. He said, "They spoke all languages and were very clear and easily understood."[12] The requirement of clarity is in the orthogonal complement of my earlier definition of truthiness. Telling the truth is of no help if the impression it leaves is fuzzy or, worse, incorrect. The media tend to eschew displays of great clarity like scatter plots, because they are thought to be too dreary or too technical. The scientific community may avoid clear displays because individual scientists lack the training to make them clear and/or the empathy to care. I view misleading your readers out of ignorance as a venial sin, because it can be fixed with training. Using truthiness to twist the conclusions of the audience toward "that which is false" just to suit your own point of view is a mortal

[9] E.g., American Society of Mechanical Engineers standards in 1914, 1938, and 1960.

[10] E.g., Bertin 1973; Tufte 1983, 1990, 1996, 2006; and Wainer 2000, 2005.

[11] This is a gentler version of Wolfgang Pauli's well-known quip that, "Science advances one funeral at a time."

[12] Reported in William Playfair (1822–3) in an unpublished manuscript held by John Lawrence Playfair, Toronto, Canada (transcribed and annotated by Ian Spence).

sin. As we saw in the discussion of the effects of fracking and wastewater injection on earthquakes in Oklahoma (Chapter 6), clarity was never an issue; quite the opposite, the goal was to obfuscate the connection. I fear that by showing some ways that data displays can confuse I may be inadvertently aiding those who are enemies of the truth. I hope not.

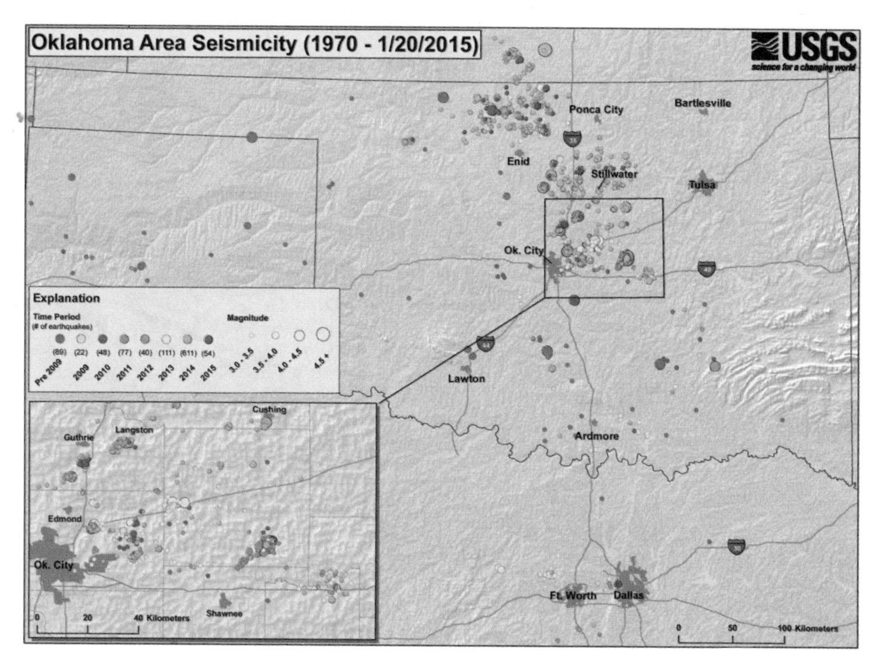


FIGURE 6.2. The geographic distribution of 3.0+ earthquakes in Oklahoma since 1970. The clusters surround injection wells for the disposal of wastewater (from the USGS).

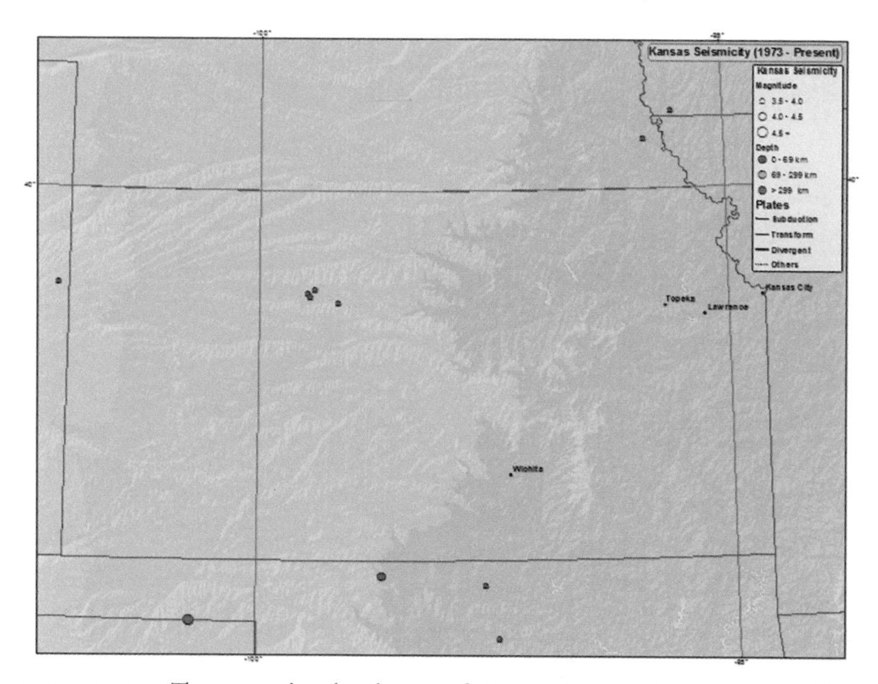


FIGURE 6.3. The geographic distribution of 3.5+ earthquakes in Kansas since 1973 (from the USGS).

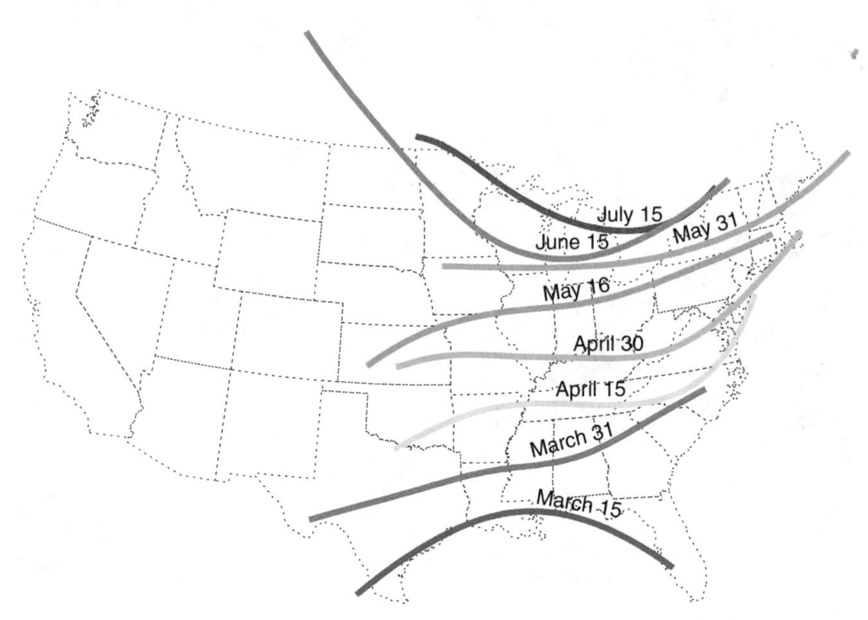


FIGURE C.1. Migration pattern of Monarch butterflies returning north after wintering in Mexico.

10

# Inside Out Plots

The modern world is full of complexity. Data that describe it too often must mirror that complexity. Statistical problems with only one independent variable and a single dependent variable are usually only found in textbooks. The real world is hopelessly multivariate and filled with interconnected variables. Any data display that fails to represent those complexities risks misleading us. Einstein's advice that "everything should be as simple as possible, but no simpler" looms prescient.

If we couple Einstein's advice with Tukey's (1977) observation (discussed in the introduction to this section) that the best way to find what we are not expecting is with a well-designed graphic display, we have an immediate problem. Most of our data displays must be represented on a two-dimensional plane. Yet trying to show three or four or more dimensions on a two-dimensional surface requires something different than the usual metaphor of representing the data spatially, for example, bigger numbers are represented by a bigger bar, a larger pie segment, a line that reaches higher, or any of the other Cartesian representations.

Happily there have been large numbers of ingenious methods developed to display multivariate data on a two-dimensional surface.[1]

[1] These include:

(1) Icons that contain many features each paired with a variable whose size or shape relate to the size of the variable – e.g., polygons or cartoon faces.
(2) Complex periodic functions where each variable represented is paired with a separate Fourier component.

(*this note is continued on the next page*)

Forty years ago Yale's John Hartigan proposed a simple approach for looking at some kinds of multivariate data. This is now called the "Inside Out Plot." Most data begin as a table, and so it is logical that we use a semigraphic display to help us look at such tabular data. The construction of an inside out plot builds on the idea that sometimes a well-constructed table can be an effective display, giving us hope that a mutated table can help us look at high-dimensional data within the limitations of a two-dimensional plotting surface.

As in most instances on the topic of display, explanation is best done through the use of an example. The popularity of the movie *Moneyball* provides the topic.

## A Multivariate Example: Joe Mauer vs. Some Immortals

In the February 17, 2010 issue of *USA Today* there was an article about the Minnesota Twins all-star catcher Joe Mauer. Mauer's first six years in the major leagues have been remarkable by any measure, but especially from an offensive perspective (during that time he won three batting titles). The article's author (Bob Nightingale) tries to cement his point by comparing Mauer's offensive statistics with those of five other great catchers during their first six years. The data he presents are shown here as Table 10.1.

How can we look at these data to see what messages they carry? Obviously we would like to somehow summarize across all of these offensive categories, but how? They are a mixture of variables with different locations and different scales. Is a batting average of .327 better than batting in 102 runs? How is a raven like a writing desk? Before they can be compared, and thence summarized, we must first somehow place all of the variables on a common scale. We'll do this in two steps. First we'll center each column by subtracting out some middle value of each

(3) Minard's sublime six-dimensional map showing Napoleon's ill-fated Russian campaign as a rushing river crossing into Russia from Poland and a tiny stream trickling back after marching back from Moscow in the teeth of the Russian winter.

There are many other approaches.

TABLE 10.1. *Offensive Statistics for the First Six Years in the Careers of Six Great Catchers*

| Mauer vs. Other Catching Greats after Six Seasons | | | | | | | | | | |
|---|---|---|---|---|---|---|---|---|---|---|
| Player | Years | At Bats | Runs | Hits | Home Runs | RBI | Batting Average | On Base Percentage | Slugging Percentage | OPS |
| Joe Mauer | 2004–09 | 2,582 | 419 | 844 | 72 | 397 | 0.327 | 0.408 | 0.483 | 0.892 |
| Mickey Cochrane | 1925–30 | 2,691 | 514 | 867 | 53 | 419 | 0.322 | 0.402 | 0.472 | 0.874 |
| Yogi Berra | 1946–51 | 2,343 | 381 | 701 | 102 | 459 | 0.299 | 0.348 | 0.498 | 0.845 |
| Johnny Bench | 1967–72 | 2,887 | 421 | 781 | 154 | 512 | 0.271 | 0.344 | 0.488 | 0.822 |
| Ivan Rodriguez | 1991–06 | 2,667 | 347 | 761 | 68 | 340 | 0.285 | 0.324 | 0.429 | 0.753 |
| Mike Piazza | 1992–07 | 2,558 | 423 | 854 | 168 | 533 | 0.334 | 0.398 | 0.576 | 0.974 |

OPS = On Base + Slugging.

column and then second scale each column into a common metric. Once this is done we can make comparisons across columns and characterize the overall performance of each player. Exactly how to do this will become clearer as we go along.

But first we must clean up the table. We can begin by first simplifying it by removing the column indicating the years they played. It may be of some background interest but it is not a performance statistic. Also, because the OPS is just the sum of two other columns (on-base percentage and slugging percentage), keeping it in will just give extra weight to its component variables. There seems to be no reason to count those measures twice and so we will also elide the OPS column. In Table 10.2 this shrunken table augments each column by the median value for that variable. This augmentation allows us to easily answer the obvious question "what's a typical value for this variable?" We choose the median, rather than the mean, because we want a robust measure that will not be overly affected by an unusual data point, and that, once removed, will allow unusual points to stick out. Also, because it is merely the middle value of those presented, it is very easy to compute.

Now that the table is cleaned up, we can center all of the columns by subtracting out the column medians. Such a table of column-centered variables is shown as Table 10.3. After the columns are all centered we see that there is substantial variability within each column. For example, we see that Ivan Rodriguez had ninety-nine fewer RBIs than the median and Mike Piazza had ninety-four more than the median, a range of 193. But compare this to batting averages in which Johnny Bench's was .040 lower than the median whereas Mike Piazza's was .024 above the median, a difference of .064. How are we to compare .064 batting average points with 193 RBIs? It is that raven and writing desk again. Obviously, to make comparisons we need to equalize the variation within each column. This is easily done by characterizing the amount of variability in each column and then dividing all elements of that column by that characterization. But how?

At the bottom of each column is the Median Absolute Deviation (MAD). This is the median of the absolute values of all of the entries in

TABLE 10.2. *Original Data with Years and OPS Elided and Column Medians Calculated*

| Player | At Bats | Runs | Hits | Home Runs | RBI | Batting Average | On Base Percentage | Slugging Percentage |
|---|---|---|---|---|---|---|---|---|
| Joe Mauer | 2,582 | 419 | 844 | 72 | 397 | 0.327 | 0.408 | 0.483 |
| Mickey Cochrane | 2,691 | 514 | 867 | 53 | 419 | 0.322 | 0.402 | 0.472 |
| Yogi Berra | 2,343 | 381 | 701 | 102 | 459 | 0.299 | 0.348 | 0.498 |
| Johnny Bench | 2,887 | 421 | 781 | 154 | 512 | 0.271 | 0.344 | 0.488 |
| Ivan Rodriguez | 2,667 | 347 | 761 | 68 | 340 | 0.285 | 0.324 | 0.429 |
| Mike Piazza | 2,558 | 423 | 854 | 168 | 533 | 0.334 | 0.398 | 0.576 |
| **Median** | **2,625** | **420** | **813** | **87** | **439** | **0.311** | **0.373** | **0.486** |

TABLE 10.3. *Results from Table 10.2 Column Centered by Subtracting Out Column Medians*

| | Column Centered | | | | | | | |
|---|---|---|---|---|---|---|---|---|
| Player | At Bats | Runs | Hits | Home Runs | RBI | Batting Average | On Base Percentage | Slugging Percentage |
| Joe Mauer | -43 | -1 | 32 | -15 | -42 | 0.017 | 0.035 | -0.003 |
| Mickey Cochrane | 67 | 94 | 55 | -34 | -20 | 0.012 | 0.029 | -0.014 |
| Yogi Berra | -282 | -39 | -112 | 15 | 20 | -0.012 | -0.025 | 0.013 |
| Johnny Bench | 263 | 1 | -32 | 67 | 73 | -0.040 | -0.029 | 0.003 |
| Ivan Rodriguez | 43 | -73 | -52 | -19 | -99 | -0.026 | -0.049 | -0.057 |
| Mike Piazza | -67 | 3 | 42 | 81 | 94 | 0.024 | 0.025 | 0.090 |
| **MAD** | **67** | **21** | **47** | **27** | **58** | **0.020** | **0.029** | **0.013** |

Median Absolute Deviations (MADs) calculated.

that column. The MAD is a robust measure of spread.[2] We use the MAD instead of the standard deviation for exactly the same reasons that we used the median.

Now that we have a robust measure of scale we can divide every column by its MAD.[3] This finally allows us to summarize each player's performance across columns. We do this by taking row medians, which, because all entries are centered, we can interpret as the player's offensive value – what we can call the player effects.[4] This is shown in Table 10.4.

The player "effects" provide at least part of the answer to the questions for which these data were gathered. We can see the outcome more easily if we reorder the players by the player effects. Such a reordered table is shown as Table 10.5.

Now we can see that Mike Piazza is, overall, the best performing offensive catcher among this illustrious group, and Ivan Rodriguez the worst. We also see that Joe Mauer clearly belongs in this company, sitting comfortably between Johnny Bench and Yogi Berra.

Of course, we can get an immediate visual sense of the size of the differences between these catchers if we display the player effects as a stem-and-leaf display[5] (see Figure 10.1).

But this is only part of the message contained in this data set. A second issue, at least as important as the overall rankings, is an understanding of any unusual performances of these players on one or more of the component variables. To understand these we must first remove the row effects by subtracting them out, and look at the residuals that remain. This result is shown in Table 10.6. On its right flank are the player effects, and its entries are the doubly centered and column-scaled residuals. If we want to see the extent to which a specific player does unusually well or unusually poorly on one or more of the various performance measures it will be worth our time to look at these residuals carefully. But how?

[2] Indeed, when the data are Gaussian, it is, in expectation, a fixed fraction of the standard deviation.

[3] Thus each column is centered and scaled as a kind of robust z-score.

[4] In traditional statistical jargon these are the *column standardized row effects*.

[5] The stem-and-leaf display was proposed by John Tukey as a simple display to show the distribution of a single variable quickly to the flying eye. The stem is just a vertically arranged, equally spaced, list of the numbers under consideration. The leaves are labels associated with those numbers.

TABLE 10.4. *Results from Table 10.3 Rescaled by Dividing Each Entry in a Column by its MAD and Row Medians (Player Effects) Calculated*

| | Column Standardized | | | | | | | | |
|---|---|---|---|---|---|---|---|---|---|
| Player | At Bats | Runs | Hits | Home Runs | RBI | Batting Average | On Base Percentage | Slugging Percentage | **Player Effects** |
| Joe Mauer | −0.64 | −0.05 | 0.68 | −0.57 | −0.73 | 0.83 | 1.21 | −0.19 | **−0.12** |
| Mickey Cochrane | 1.00 | 4.48 | 1.17 | −1.28 | −0.35 | 0.58 | 1.00 | −1.04 | **0.79** |
| Yogi Berra | −4.23 | −1.86 | −2.40 | 0.57 | 0.35 | −0.58 | −0.86 | 0.96 | **−0.72** |
| Johnny Bench | 3.95 | 0.05 | −0.68 | 2.53 | 1.27 | −1.98 | −1.00 | 0.19 | **0.12** |
| Ivan Rodriguez | 0.64 | −3.48 | −1.11 | −0.72 | −1.72 | −1.28 | −1.69 | −4.35 | **−1.48** |
| Mike Piazza | −1.00 | 0.14 | 0.89 | 3.06 | 1.63 | 1.18 | 0.86 | 6.96 | **1.03** |

TABLE 10.5. *Table 10.4 with its Rows Reordered by Row (Player) Effects*

| | Column Standardized and Reordered | | | | | | | | |
|---|---|---|---|---|---|---|---|---|---|
| Player | At Bats | Runs | Hits | Home Runs | RBI | Batting Average | On Base Percentage | Slugging Percentage | **Player Effects** |
| Mike Piazza | -1.00 | 0.14 | 0.89 | 3.06 | 1.63 | 1.18 | 0.86 | 6.96 | **1.03** |
| Mickey Cochrane | 1.00 | 4.48 | 1.17 | -1.28 | -0.35 | 0.58 | 1.00 | -1.04 | **0.79** |
| Johnny Bench | 3.95 | 0.05 | -0.68 | 2.53 | 1.27 | -1.98 | -1.00 | 0.19 | **0.12** |
| Joe Mauer | -0.64 | -0.05 | 0.68 | -0.57 | -0.73 | 0.83 | 1.21 | -0.19 | **-0.12** |
| Yogi Berra | -4.23 | -1.86 | -2.40 | 0.57 | 0.35 | -0.58 | -0.86 | 0.96 | **-0.72** |
| Ivan Rodriguez | 0.64 | -3.48 | -1.11 | -0.72 | -1.72 | -1.28 | -1.69 | -4.35 | **-1.48** |

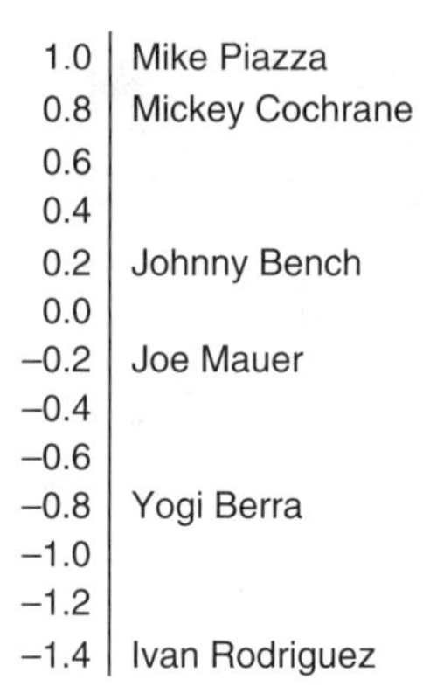


FIGURE 10.1. Stem-and-leaf diagram of player effects.

The format of Table 10.6 is completely standard. It has labels on the outside of the table and numbers on the inside. To view the results in this table directly – and to get to the point of this chapter – we must turn the table inside out and put the numbers on the outside and the labels on the inside.

Such an inside out plot is shown in Figure 10.2. Note we used the convention of using the plotting symbol /–/ to represent all the unnamed players whose residuals were essentially zero on that variable, when there are too many to fit their names in explicitly. This is sensible because it renders anonymous those players whose residuals on this variable are too small to be of interest anyway.

Even a quick glance at Figure 10.2 provides enlightenment. We see that Johnny Bench had an unusually high number of at bats, whereas Yogi Berra and Mike Piazza had fewer than we would expect from their overall ranking. Mickey Cochran scored a great deal more runs than we would have expected, although many fewer home runs. They were all pretty consistent RBI producers. Joe Mauer seems to stand out on three highly related variables: hits, batting average, and on-base percentage, whereas Johnny Bench is at or near the opposite extreme on those. And the biggest residual is reserved for Mike Piazza's slugging percentage, which outshines even Yogi Berra on this measure.

The inside out plot provides a simple robust way to look at data that are not measured on the same scale. We do not suggest that other approaches will not also bear fruit, but only that this one is simple and

TABLE 10.6. *The Results from Table 10.5 with the Row Effects Removed. A Doubly Centered and Column Scaled Data Matrix*

| | Column Standardized and Reordered with Row Effects Removed | | | | | | | | |
|---|---|---|---|---|---|---|---|---|---|
| Player | At Bats | Runs | Hits | Home Runs | RBI | Batting Average | On Base Percentage | Slugging Percentage | **Player Effects** |
| Mike Piazza | −2.03 | −0.89 | −0.14 | 2.03 | 0.60 | 0.15 | −0.17 | 5.93 | **1.03** |
| Mickey Cochrane | 0.21 | 3.69 | 0.38 | −2.07 | −1.14 | −0.21 | 0.21 | −1.83 | **0.79** |
| Johnny Bench | 3.83 | −0.07 | −0.80 | 2.41 | 1.15 | −2.10 | −1.12 | 0.07 | **0.12** |
| Joe Mauer | −0.52 | 0.07 | 0.80 | −0.45 | −0.61 | 0.95 | 1.33 | −0.07 | **−0.12** |
| Yogi Berra | −3.51 | −1.14 | −1.68 | 1.29 | 1.07 | 0.14 | −0.14 | 1.68 | **−0.72** |
| Ivan Rodriguez | 2.12 | −2.00 | 0.37 | 0.76 | −0.24 | 0.20 | −0.21 | −2.87 | **−1.48** |

|  | At Bats | Runs | Hits | Home Runs | RBI | Batting Average | On Base Percentage | Slugging Percentage |
|---|---|---|---|---|---|---|---|---|
| 6 |  |  |  |  |  |  |  | Piazza |
| 5 |  |  |  |  |  |  |  |  |
| 4 | Johnny Bench | Cochrane |  |  |  |  |  |  |
| 3 |  |  |  |  |  |  |  |  |
| 2 | Ivan Rodriguez |  |  | Piazza, Bench | Bench, Berra |  |  | Berra |
| 1 |  |  | Mauer | Berra, Rodriguez | Piazza | Mauer | Mauer |  |
| 0 | Mickey Cochrane | Mauer, Bench | — | Mauer | Rodriguez | — | — | — |
| –1 | Joe Mauer | Piazza, Berra | Bench |  | Mauer, Cochrane |  | Bench |  |
| –2 | Mike Piazza | Rodriguez | Berra | Cochrane |  | Bench |  | Cochrane |
| –3 | Yogi Berra |  |  |  |  |  |  | Rodriguez |

FIGURE 10.2. Standardized residuals plotted inside out.

easy – providing a way to look at the fit and also at the residuals from the fit. It requires little more special processing beyond that provided by a spreadsheet.

Of course, with a toy example like this, comprised of only six players and eight variables, much of what we found is also seen through the careful study of the component tables (e.g., Table 10.6). The value of inside out plotting would be clearer with a larger sample of players. If it included competent but not all-star catchers it would have made the case of the specialness of Joe Mauer clearer. It might also help to look at a parallel set of defensive statistics. We suspect that, had this been done, Ivan Rodriguez would have shone in a brighter light. Although larger data sets would have made the power of this technique more obvious, it would also have been more cumbersome; this was the suitable size for a demonstration.

But how far does this approach scale upward? Suppose we had twenty or eighty catchers? Inside out plots will still work, although some modifications may be helpful; replacing each catcher's name with a less evocative, but more compact representation is a start. Next, remembering that residuals near zero is the place where the most names will pile up and are also of least interest. Hence replacing large numbers of names with a single /–/ works fine.

A second issue – that of highly correlated variables – has already been hinted at with our dismissal of OPS as redundant. Decisions about which variables to include and which to elide must be made, but are not directly related to the plotting method. Although if we include two variables that are highly related the inside out plot will tell us, for the residuals will appear very similar. As is true for all investigations, inside out plots are often iterative, where one plot provides information on how to make the next one.

One can also easily imagine dynamic augmentations to inside out plots. For example, one might prepare a program for inside out plots so that if you point to a player's name a line appears that connects that name across all variables. One strategy might be to construct in a series of these augmented versions dynamically, and subsequently choose a few especially interesting ones to yield a series of static displays. We leave it to the readers' imaginations to devise other variations.

11

# A Century and a Half of Moral Statistics

## *Plotting Evidence to Affect Social Policy*

> *It is a truth universally acknowledged, that any person in possession of a geographic data set, must be in want of a map.*
>
> Jane Austen 1817

The sophisticated use of thematic maps began on November 30, 1826 when Charles Dupin (1784–1873) gave an address on the topic of popular education and its relation to French prosperity. He used a map shaded according to the proportion of male children in school relative to the size of the population in that départment. This graphic approach was improved upon in 1830 when Frére de Montizon produced a map showing the population of France in which he represented the population by dots, in which each dot represented ten thousand people. Although no one remarked on it for the better part of a century, this was perhaps the most important conceptual breakthrough in thematic mapping. It was the direct linear antecedent of John Snow's famous map of the 1854 cholera epidemic in London. Snow's map, presented in Figure 11.1, uses bars to show the location and number of cholera deaths, which Snow noted were centered around the Broad Street water pump. Even though the vectors of cholera contagion were unknown at that time,[1] the spatial

[1] Ironically, the same year as the London epidemic, the Italian Filippo Pacini identified the bacteria *Vibrio cholera* as the proximal cause of the disease.

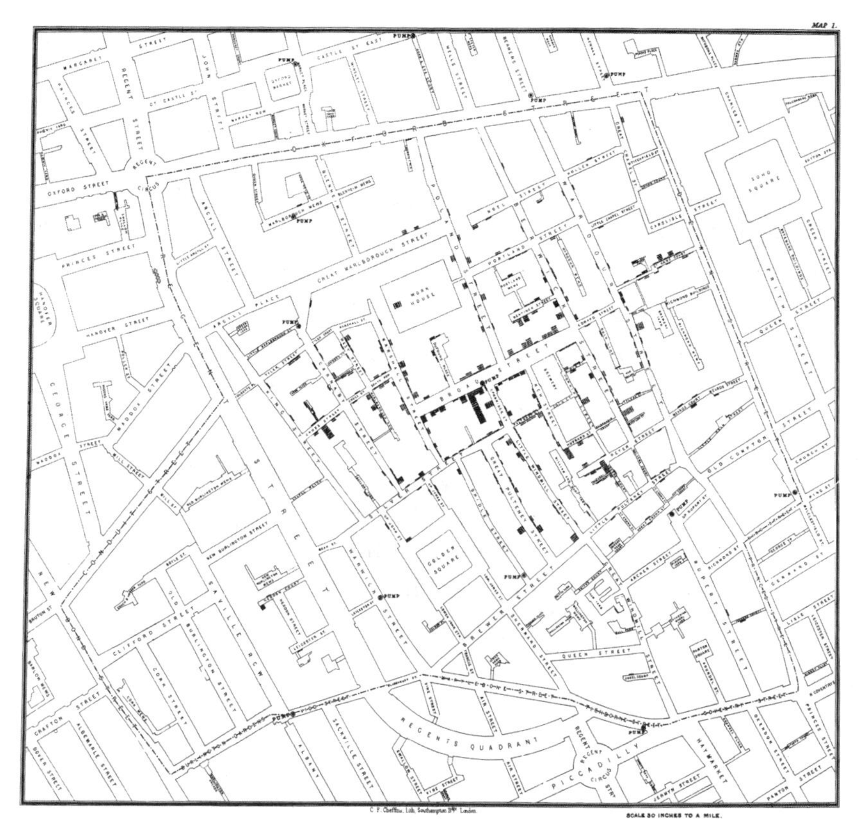


FIGURE 11.1. John Snow's 1854 cholera map of London.

pattern of deaths relative to the water pump suggested to Snow the cause of the epidemic. He petitioned the Vestry of St. James to remove the pump's handle, and within a week the epidemic, which had taken 570 lives, ended. This may not be history's most powerful example of Tukey's observation about a graph being the best way to find what you were not expecting, but it is certainly among the top five.

## Joseph Fletcher and Maps of Moral Statistics

Maps of moral statistics started appearing at about the same time; these typically focused on various aspects of crime. Most widely known were those of Adriano Balbi (1782–1848) and Andre-Michel Guerry (1802–66), whose 1829 map pairing the popularity of instruction with the incidence of crimes was but the start of their continuing investigations. Their work was paralleled by the Belgian Adolphe Quetelet (1796–1874), who, in 1831, produced a marvelous map of crimes against property in France, in which the shading was continuous across internal boundaries rather than being uniform within department. Guerry expanded his investigation in the 1830s to add three other moral variables (illegitimacy, charity, and suicides) to crime and instruction. Thus, despite graphical display's English birth, its early childhood was spent on the continent.

Maps of moral statistics began their return across the channel in 1847 when a thirty-four-year-old barrister named Joseph Fletcher (1813–52) published the first of several long papers. This paper was filled with tables and but a single, rudimentary map. In fact, he explicitly rejected the use of maps, suggesting that you could get all you needed by simply perusing the columns of numbers and thus avoid the bother and expense of drafting and printing maps. He reversed himself completely[2] two years later in two articles of the same title (but of much greater length) as his 1847 paper. These included many shaded maps that were innovative in both content and format. Format first.

Fletcher, having a background in statistics, or at least what was thought of as statistics in the mid-nineteenth century, did not plot the raw numbers but instead their deviations from the mean. And so the scale of tints varied from the middle, an approach that is common now. In 1849, the approach was an innovation. The advantage of centering all variables at zero was that it allowed Fletcher to prepare a number of maps, on very different topics and with very different scales, and place them side by side for comparison without worrying about the location of the scale. He also oriented the shading, as much as possible, in the same way. In his words,

> In all the Maps it will be observed that the *darker* tints and the *lower* numbers are appropriated to the *unfavorable* end of the scale. (158; author emphasis)

[2] Ordinarily it would be hard to know exactly what got Fletcher to change his approach so completely, but he gave an evocative hint in his 1849 paper when he stated, "A set of shaded maps accompanies these tables, to illustrate the most important branches of the investigation, and I have endeavoured to supply the deficiency which H. R. H. Prince Albert was pleased to point out, of the want of more illustrations of this kind."

Of course, with variables like population density it is hard to know which end is favorable. He chose lesser population as more favorable because, we speculate, it seemed to accompany favorable outcomes on the other variables. Although many of Fletcher's innovations in format were noteworthy, the content he chose makes him special. He did not make maps of obvious physical variables like wind direction or altitude or even the distribution of the population (although he did make one map of that, it was for comparative purposes). No, he had more profound purposes in mind. Joseph Fletcher made maps of moral statistics, and opposed these maps with others to suggest causal interpretations.

For example, beside his plot of ignorance in England and Wales (Figure 11.2) he also had a map of the incidence of crime (Figure 11.3). He opted for this juxtaposition on the basis of detailed analysis of subsets of the data. For example, he wrote,

> We thus find that the decline in *total* ignorance to be slowest in the most criminal and the most ignorant districts. (320)

He then tied this analysis to the phenomenon made observable through the juxtaposed maps, stating,

> [T]he darkest tints of ignorance go with those of crime, from the more southern of the Midland Manufacturing Counties, through the South Midland and Eastern Agricultural Counties, ... and it will be well to observe, as an example of their use, that all four of the tests of moral influences now employed are seen to be on the side of the more instructed districts.

And then, looking into the phenomenon at a more microscopic level, he observed,

> The two least criminal regions are at the opposite extremes in this respect (the Celtic and the Scandinavian), with this important difference, that in the region where there is the greatest decline of absolute ignorance among the criminals (the Scandinavian), there is not one-half of the amount of it in the population at large which exists in the other.

In addition to these maps, he also produced parallel plots of bastardy in England and Wales, of improvident marriages, of persons of

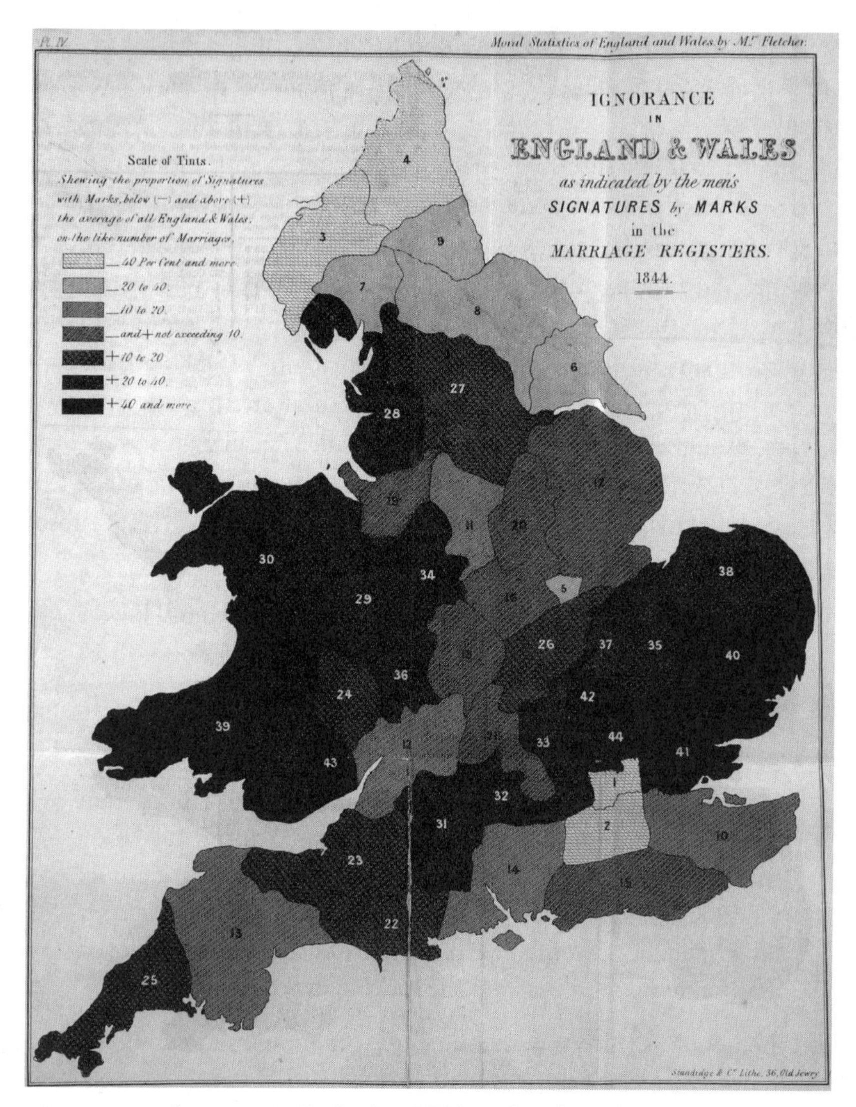


FIGURE 11.2. Ignorance in England and Wales, taken from Fletcher (1849b).

independent means, of pauperism, and of many other variables that could plausibly be thought of as either causes or effects of other variables. His goal was to generate and test hypotheses, which might then be used to guide subsequent social action and policy.

Though some understanding of the relation between crime and ignorance, as well as their relation over time, can be attained by comparing

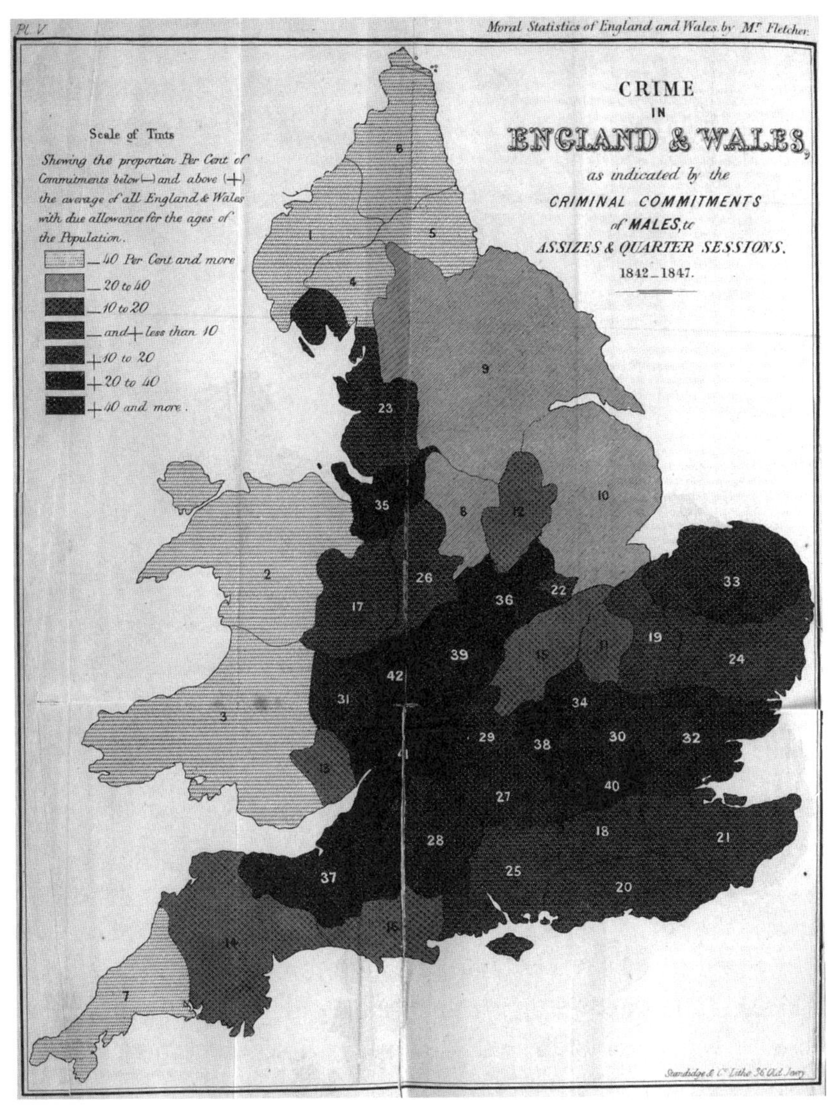


FIGURE 11.3. Crime in England and Wales, taken from Fletcher (1849b).

thematic maps, the process is neither easy nor precise – even using modern analysis and shading techniques. Fletcher's use of thematic maps was innovative but perhaps not the ideal tool. Had he been interested in astronomy, Fletcher might have come upon a more useful tool for depicting the relation between these two data sets: the scatter plot. In his "investigation of the orbits of revolving double stars," published in

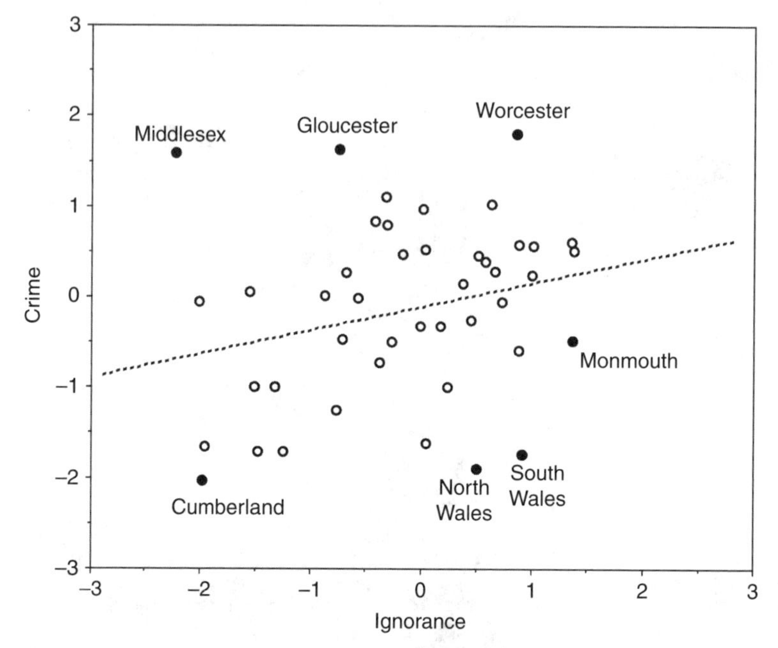


FIGURE 11.4. A scatter plot showing the modest relation between ignorance and crime in Fletcher's nineteenth-century England, with outlying counties filled in and identified. The metric used is the number of standard deviations from the median.

1833, British astronomer John Frederick William Herschel plotted the positions of stars on one axis against the observation year on the other and drew a smoothed curve between the points to represent the relation between the two variables and, in so doing, unknowingly staked his claim as inventor of what is now called a *scatter plot*.[3] Such a representation of Fletcher's 1845 crime and ignorance data more clearly illustrates the modest positive relation between the two variables and is presented here in Figure 11.4.

## Guns, Murders, Life, Death, and Ignorance in Contemporary America

Fletcher's principal goal in preparing his essay was to affect opinion on what he considered to be key social issues. Fletcher drew a causal

[3] Friendly and Denis 2005.

connection suggesting that improving schooling would simultaneously reduce ignorance and crime. A reanalysis of the same data a century later showed that the policies of improved public education favored by Fletcher had the effect that he anticipated.

Though we firmly believe in a causal connection between ignorance and crime, empirical evidence to support such a claim, even partially, is always welcome. Some relations between various contemporary moral statistics seem, on their face, to be obvious, yet stakeholders of one sort or another dispute such claims, regardless of how obvious they seem. So, let us take a page from Fletcher's book and look at evidence for a variety of such claims.

I shall make two kinds of claims and hence mobilize two kinds of displays in their support. The first kind of claim will be about the relation between two quantitative variables, for example the relation between the number of guns in a state and the number of people murdered by guns in that state. These claims will be illustrated with scatter plots. Such plots show the relation between the variables but convey little geographic information. We augment the scatter plots a little by indicating the states that voted in the majority for Barack Obama in the 2012 presidential election ("blue states") and those that voted for Mitt Romney ("red states")

The second kind of claim relates more directly to the geographic distribution of the variables displayed in the scatter plots. To accomplish this, we shall use the same choropleth format that Fletcher employed more than 150 years ago. We modernize his format slightly, following the lead of Andrew Gelman[4] in his maps of the 2008 presidential election. We shade each map so that states above the median in the direction generally agreed as positive (e.g., fewer murders, longer life expectancy, lesser ignorance, greater income) will be in blue (the further from the median, the more saturated the blue). And states below the median are shaded in red (the lower, the more saturated the red). The closer to the median that a state scores, the paler is the coloration, so that a state at the median is shaded white. Our goal is to examine the extent to which geography is destiny. Is there coherence among states that always seem to find themselves at the unfortunate end of these variables? Perhaps such a view can

[4] Gelman 2008.

suggest ways that those states can amend their policies to improve the lives of their inhabitants. Or, if those states resist such change, the maps can guide their inhabitants to alternative places to live.

## Variables and Data Sources

**Variable 1.** Number of guns in a state – we found it difficult to get an accurate estimate, so we used, as a proxy, the 2012 NCIS firearm background checks per one hundred thousand residents in each state. We assume that the more background checks, the more guns in the state. We assume that the actual number of guns purchased far exceeds the number of background checks, but we expect that the relation between the two is monotonic. Some might claim that by not requiring background checks we might thus reduce gun violence, but we remain unconvinced by this argument.

**Variable 2.** Firearm death rates per one hundred thousand residents from statehealthfacts.org.

**Variable 3.** 2010–11 life expectancy from the American Human Development Project.

**Variable 4.** Ignorance. We used the 2011 eighth grade reading score from NAEP (popularly referred to as "The Nation's Report Card") as a measure of the ignorance of the state's population. NAEP scores are a random sample from the population, and all scores on all age groups and all subjects are strongly and positively related to one another,[5] and hence we can simply chose any one and it would be representative of all of them. We chose eighth grade reading, but any other would have yielded essentially the same picture. Fletcher used the proportion of individuals who signed their wedding license with an "X" as his indicant of ignorance. We believe that our choice represents an improvement.

**Variable 5.** Income. 2010 per capita income from U.S. Census Bureau *2012 Statistical Abstracts.*

[5] More technically, they form a "tight positive manifold" in the exam space.

## Claims about the Relation between Variables

Claim 1. The more people killed by guns in a state, the lower life expectancy will be in that state (see Figure 11.5).
Claim 2. The more guns registered in a state, the greater will be the number of gun deaths (see Figure 11.6).
Claim 3. The greater the ignorance in a state, the greater the number of gun deaths (see Figure 11.7).
Claim 4. The lesser the ignorance, the greater the income (see Figure 11.8).

## Geographic Claim

There is a coherence among states that has remained stable for at least the past eighty years. In 1931, Angoff and Mencken tried to determine which was the worst American state. Their conclusions, based on a large mixture of indices of quality of life, closely resemble what

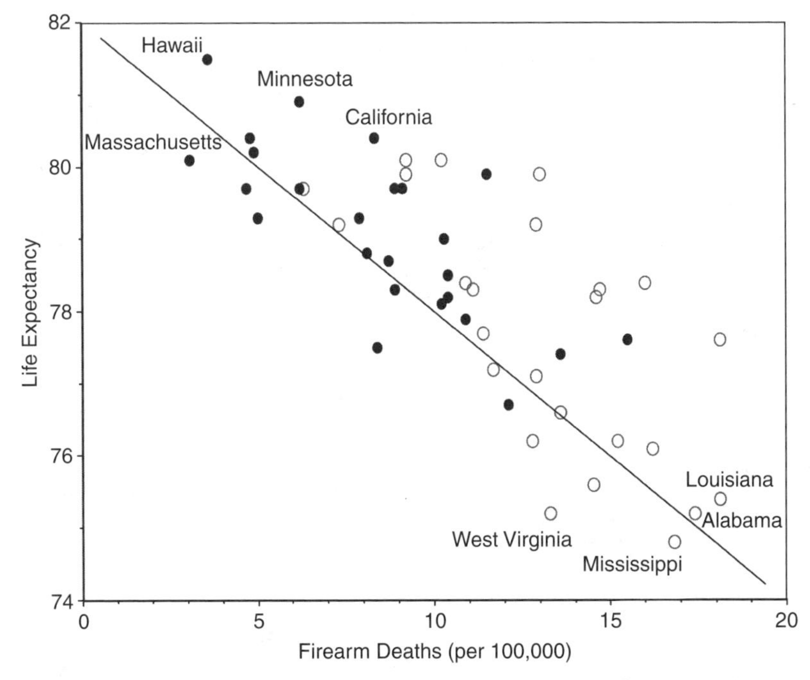


FIGURE 11.5. A scatter plot of 2010–11 life expectancy versus firearm death rate per one hundred thousand by state. The solid dots are states that voted for Obama in the 2012 presidential election; states with the open circles voted for Romney.

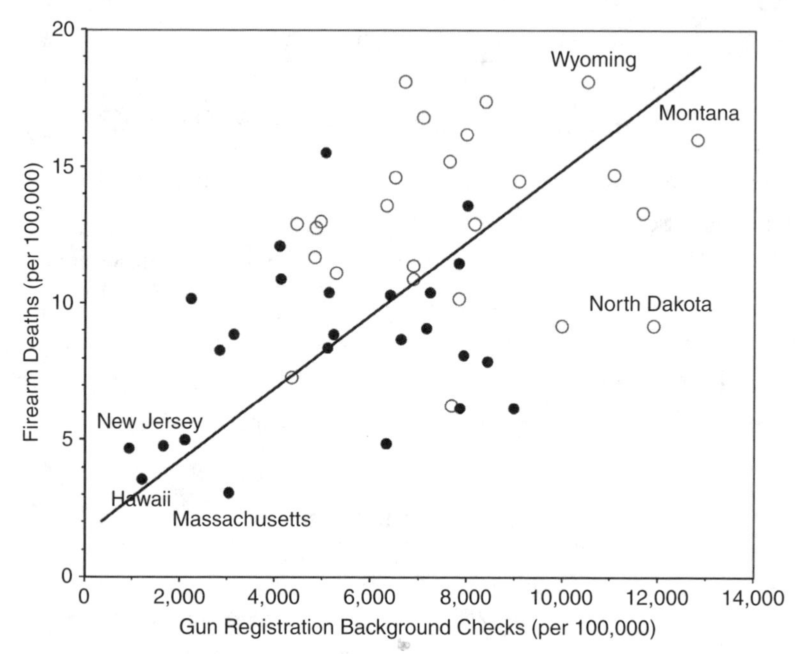


FIGURE 11.6. The horizontal axis shows the 2012 NCIS firearm background checks per one hundred thousand in each state, the vertical axis is the firearm death rate per one hundred thousand. Once again, states with solid dots are states that voted for Obama in 2012; states with open circles voted for Romney.

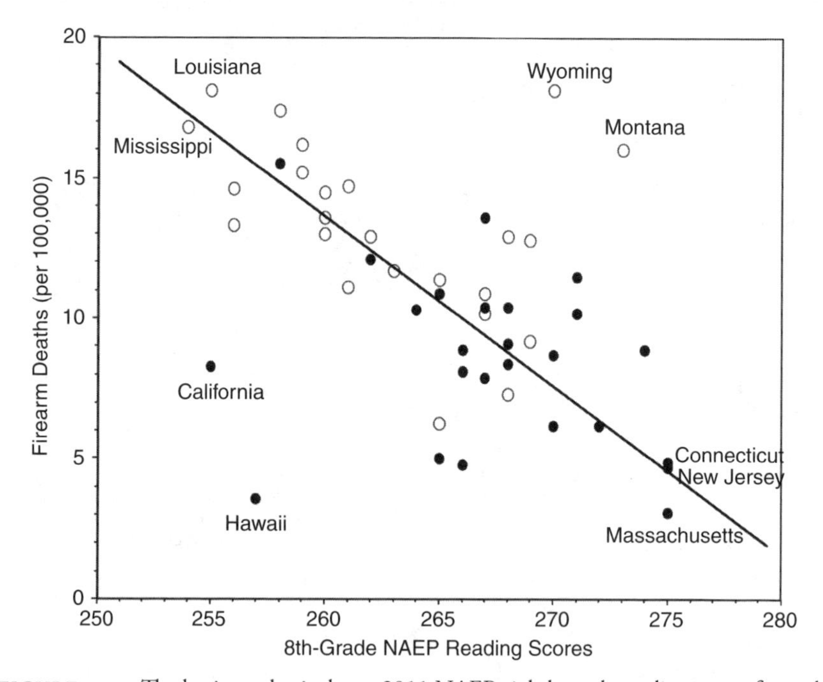


FIGURE 11.7. The horizontal axis shows 2011 NAEP eighth grade reading scores for each state; the vertical axis has the firearm death rate per one hundred thousand. Solid dots represent states that voted for Obama in 2012; states with open circles voted for Romney.

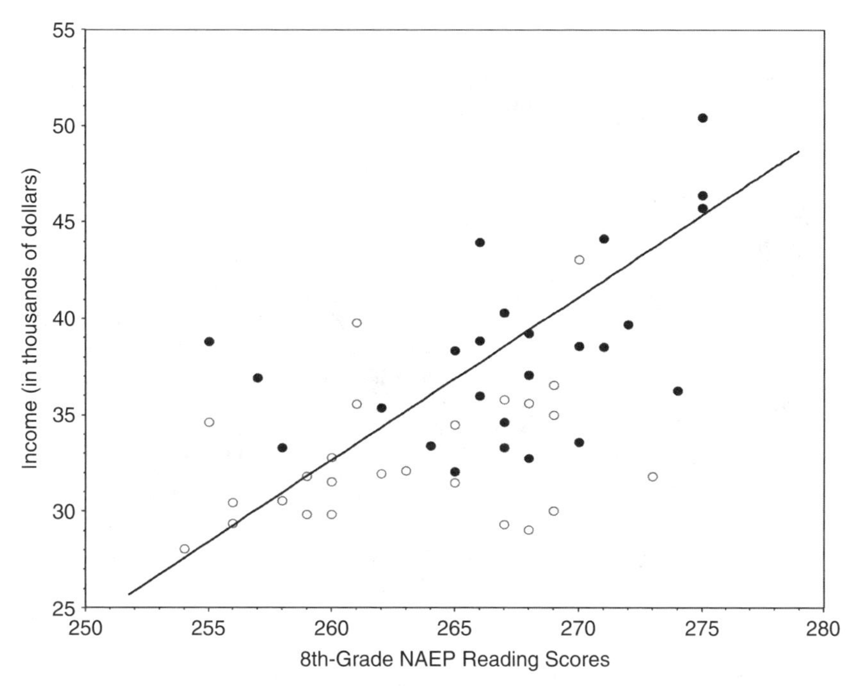


FIGURE 11.8. The horizontal axis shows 2011 NAEP eighth grade reading scores; the vertical axis shows the 2010 per capita income. The solid dots represent states that voted for Obama in 2012; the open circles voted for Romney.

we see in our twenty-first-century data. Though people differ in their goals and aspirations, most would agree that life on the whole, is better than death (Figure 11.9 – life expectancy; Figure 11.10 – firearm death rates), that some provision for human need is better than chronic lack (Figure 11.11 – income), and that a literate population is better able to participate in decisions about the future and welfare of themselves and their progeny (Figure 11.12 – NAEP Reading Scores).

Even a cursory study of these figures reveals a startling similarity in the geographic patterns of these variables. This pattern repeats itself in Figure 11.13, gun background checks, suggesting a plausible causal connection that seems worth more serious consideration. The red and blue coloration scheme used here is based solely on the states' location on the variable plotted, not on their voting in the 2012 presidential election (Figure 11.14). Yet the similarities of their color patterns to those of the voting map are striking. The direction of the causal arrow connecting the voting habits of states and their position on these variables is uncertain.

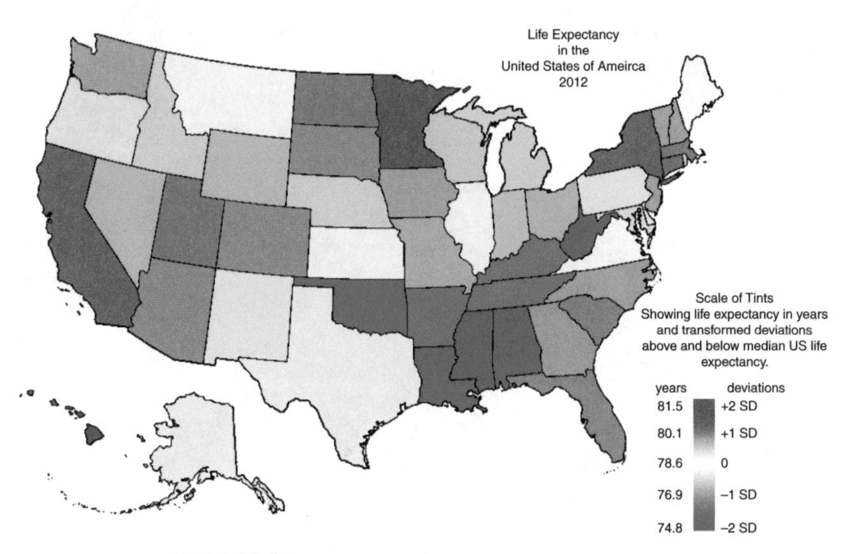


FIGURE 11.9. 2012 U.S. life expectancy.

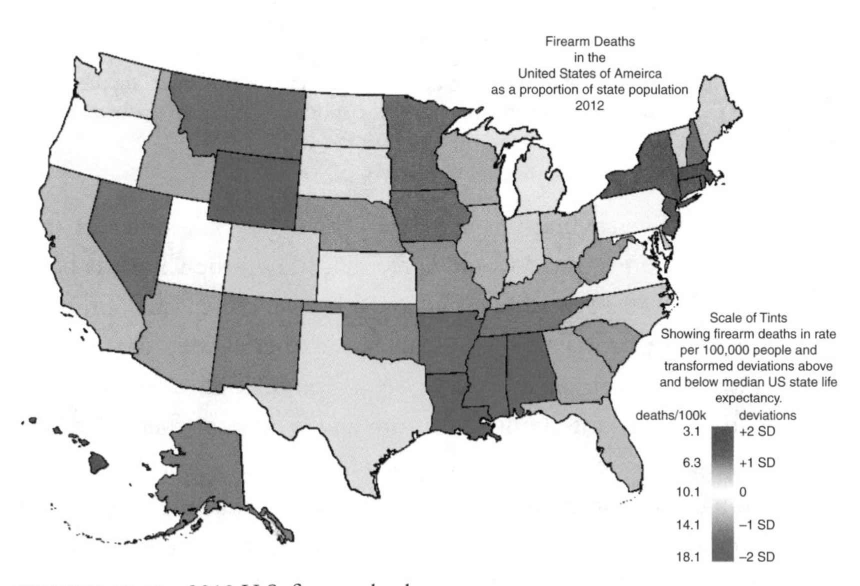


FIGURE 11.10. 2012 U.S. firearm death rate.

Are the citizens of states ignorant because of the policies espoused by Governor Romney? Or are ignorant people attracted to such policies? Our data do not illuminate this question.

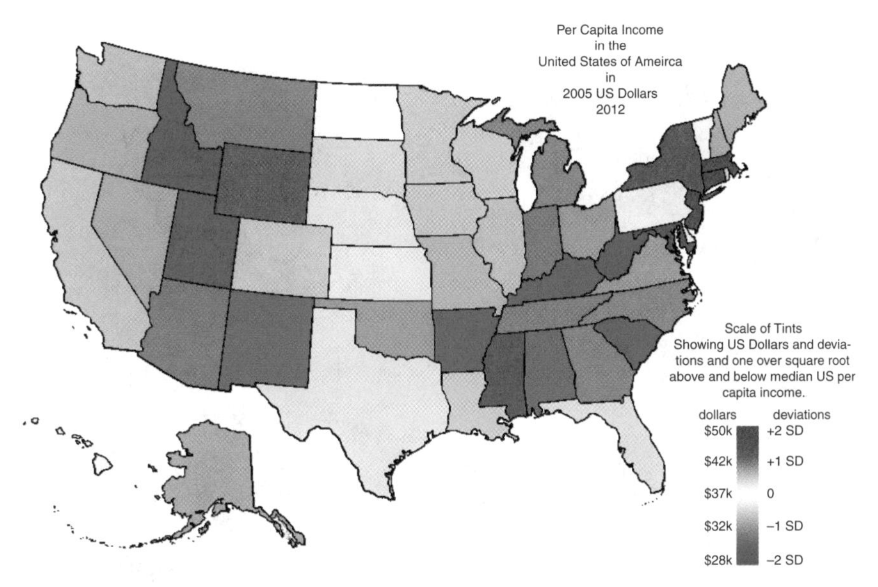


FIGURE 11.11. 2012 U.S. per capita income.

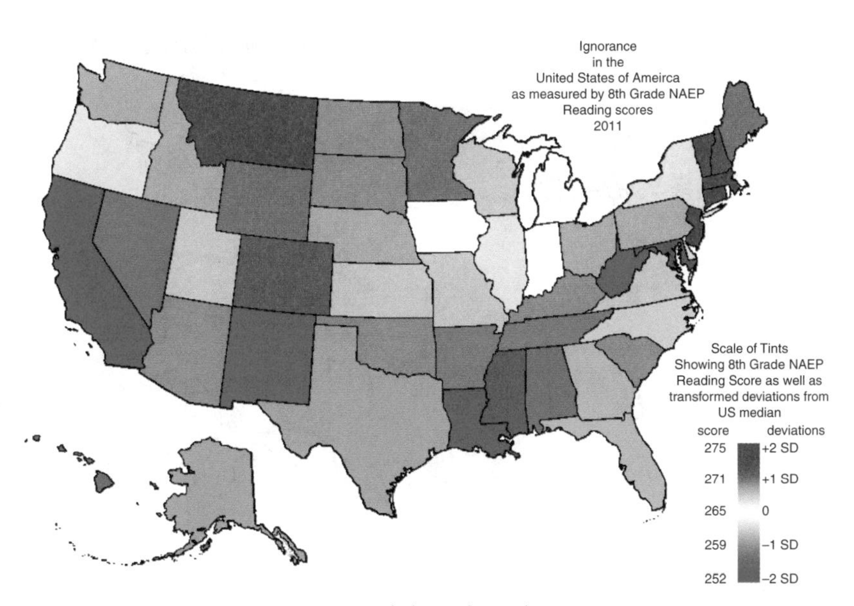


FIGURE 11.12. 2011 NAEP scores: Eighth grade reading.

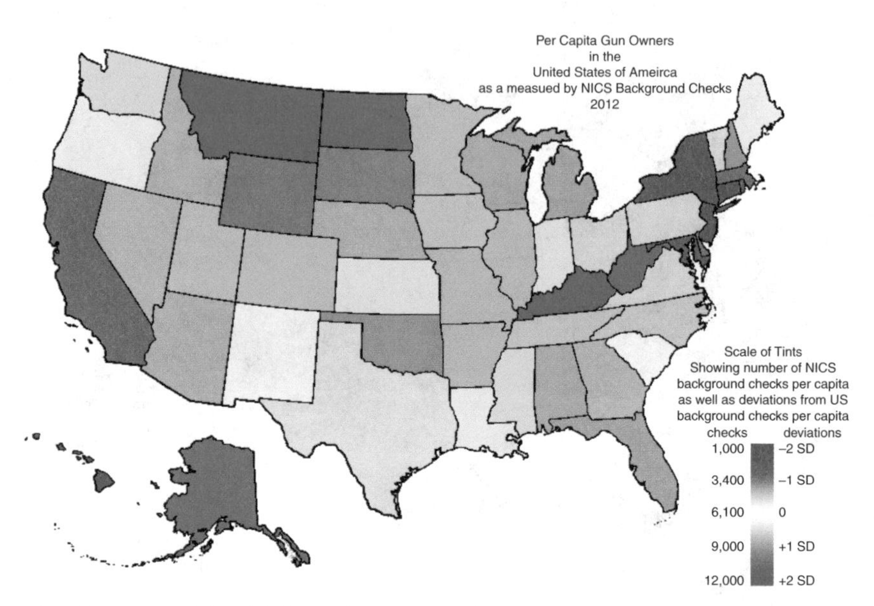


FIGURE 11.13. 2012 NCIS background checks.

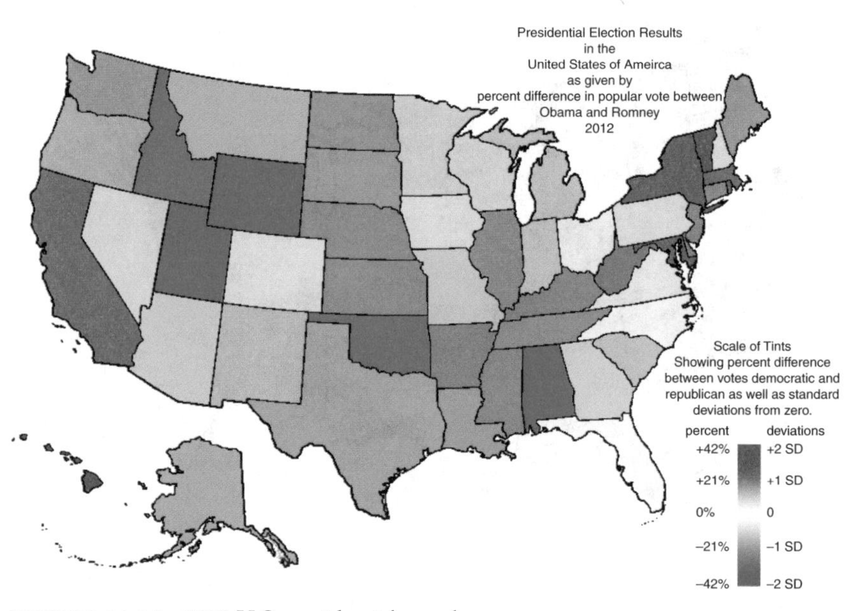


FIGURE 11.14. 2012 U.S. presidential popular vote.

## Conclusion

The investigation we have performed here is not causal in any serious way. To credibly make clausal claims we would need to hew more closely to the approach laid out in Chapter 3, using Rubin's Model. Indeed the rough-and-ready approach used here was designed to be exploratory – what might be called "reconnaissance at a distance," which suggested possible causal arrows that could be explored more deeply in other ways. Such an exploratory approach is perfectly fine so long as the investigator is well aware of its limitations. Principal among them is the ecological fallacy, in which apparent structure exists in grouped (e.g., average state) data that disappears or even reverses on the individual level. All that said, investigations like these provide circumstantial causal evidence that can often prove to be quite convincing.[6]

Joseph Fletcher's remarkable work provided a springboard toward the illumination of both the strengths and the weaknesses of choropleth maps. Their strengths are answering the dual questions "what's happening here?" and "where is this happening?" They are less powerful in helping us make comparisons between two variables. For example, are places that are ignorant also crime ridden? Questions of the latter type are answered better with scatter plots, which unfortunately do not convey the information about geographic distribution. This result drove us to the inexorable conclusion that the two kinds of displays should be used in tandem to fully illuminate the data.

To clarify how such a strategy can be used, we looked at data from the twenty-first-century United States and used them to illustrate how a combination of shaded maps and scatter plots can contribute to the beginning of an evidence-based discussion, which in the case we chose leads to an inexorable conclusion.

[6] Remembering Thoreau's observation that "Sometimes circumstantial evidence can be quite convincing, like when you find a trout in the milk." The structure found in the milky waters of twenty-first-century political discourse, when looked at carefully, does bear a startling resemblance to a fish.

Although Joseph Fletcher never quoted Thomas Hobbes[7] as the philosophical underpinning of the tack that he took, I shall. Hobbes described the natural state of mankind, without the control of some sort of central government, as a "war of every man against every man," and the result of such a circumstance as lives that are "solitary, poor, nasty, brutish, and short." We have seen that, as access to firearms becomes less and less controlled, lives become shorter, wealth diminishes, and ignorance increases. Or at least that is what the evidence shows.

[7] What Hobbes actually said was, "Whatsoever therefore is consequent to a time of Warre, where every man is Enemy to every man; the same is consequent to the time, wherein men live without other security, than what their own strength, and their own invention shall furnish them withall. In such condition, there is no place for Industry; because the fruit thereof is uncertain; and consequently no Culture of the Earth; no Navigation, nor use of the commodities that may be imported by Sea; no commodious Building; no Instruments of moving, and removing such things as require much force; no Knowledge of the face of the Earth; no account of Time; no Arts; no Letters; no Society; and which is worst of all, continuall feare, and danger of violent death; And the life of man, solitary, poore, nasty, brutish, and short."

SECTION III

# Applying the Tools of Data Science to Education

## Introduction

From 1996 until 2001 I served as an elected member of the Board of Education for the Princeton Regional Schools. Toward the end of that period a variety of expansion projects were planned that required the voters pass a $61 million bond issue. Each board member was assigned to appear in several public venues to describe the projects and try to convince those in attendance of their value so that they would agree to support the bond issue. The repayment of the bond was projected to add about $500 to the annual school taxes for the average house, which would continue for the forty years of the bond. It was my misfortune to be named as the board representative to a local organization of senior citizens.

At their meeting I was treated repeatedly to the same refrain: that they had no children in the schools, that the schools were more than good enough, and that they were living on fixed incomes and any substantial increase in taxes could constitute a hardship, which would likely continue for the rest of their lives. During all of this I wisely remained silent. Then, when a pugnacious octogenarian strode to the microphone, I feared the worst. He glared out at the gathered crowd and proclaimed, "You're all idiots." He then elaborated, "What can you add to your house for $500/year that would increase its value as much as this massive improvement to the schools? Not only that, you get the increase in your property value immediately, and you won't live long enough to pay even a small portion of the cost. You're idiots." Then he stepped down. A large number of the gray heads in the audience turned to one another and nodded in agreement. The bond issue passed overwhelmingly.

Each year, when the real estate tax bill arrives, every homeowner is reminded how expensive public education is. Yet, when the school system works well, it is money well spent, even for those residents without children in the schools. For as surely as night follows day, real estate values march in lockstep with the reputation of the local schools. Of course, the importance of education to all of us goes well beyond the money spent on it. The effects of schools touch everyone's lives in a deep and personal way – through our own experiences, our children's, and everyone we know. Thus it isn't surprising that education issues are so often featured in media reports. What is surprising is how often those reports are based on claims that lie well beyond the boggle threshold, with support relying more on truthiness than on evidence. The ubiquity of educational policies that rest on weak evidentiary bases, combined with the importance of education in all of our lives, justifies devoting a fair amount of effort to understanding such policies. In this section I discuss six important contemporary issues in education and try to see what evidence there is that can help us think about them.

A popular narrative in the mass media proclaims that public schools are failing, that each year they get worse, and that persistent race differences are building a permanent underclass. In Chapter 13 we look at data gathered over the last twenty years by the National Assessment of Educational Progress (NAEP – often called "the Nation's Report Card") to see if those data provide evidence to support these claims. Instead what emerges is clear evidence of remarkable, nationwide, improvements in student performance of all states, and that the improvement in minority performance was even greater than that of white students.

Part of that same narrative assigns causes for the nonexistent decline in educational performance. Most commonly, the blame falls on teachers, some of whom are claimed to be lazy and incompetent but protected from just treatment by intransigent unions and the power of tenure, which provides a sinecure for older and more highly paid teachers. Calls for the abolition of tenure are regularly heard, and some states are beginning to act on those calls. They are doing this in the hope of easing the removal of highly paid teachers and opening the door for new, younger, energetic, and cheaper teachers, and thus helping to control the ballooning costs. In Chapter 14 we consider the origins of teacher tenure and how one of its goals was, ironically, to control costs. We show data that illustrate unambiguously how removing tenure is likely to cause payroll costs to skyrocket.

Much of what we know of the performance of our educational system is based on student performance on tests. For test scores to be a valid indicator of performance, we must be sure that they represent the students' abilities and are not distorted though cheating. In Chapter 15 we learn of how one testing organization used faulty statistical measures to detect cheating, and managed to inflict unnecessary hardship without sufficient evidence. Moreover, they did this despite the existence of an easy and much more efficacious alternative.

We often hear about failing schools, whose students do not reach even minimal standards. In Chapter 16 we learn of a Memphis charter school whose state charter was threatened because their fifth graders scored zero on the statewide assessment. This remarkable result was obtained not because of poor performance by the students but because a state rule mandated that any student who did not take the test be automatically given a zero. This story and its outcome are the topic of this chapter.

The College Board is at least as much a political organization as a scientific one. So when it announces changes to its iconic college admissions test (the SAT), the announcement and the changes it describes garner substantial media attention. Thus, in March 2014 when, with great fanfare, we learned of three changes to the next version of the SAT, it seemed sensible to try to understand what the effect of those changes was going to be and to try to infer what led to the decision to make these changes. In Chapter 17 we do exactly this and find a surprising link to some advice given to Yale's then-president Kingman Brewster about how to smooth Yale's passage to admitting women.

In 2014 the media was full of reports decrying the overuse of tests in the United States. A Google search of the phrase "Testing Overload in America's Schools" generated more than a million hits. Is too much time spent testing? In Chapter 18 we examine the evidence surrounding the actual length of tests and try to understand how much is enough. The conclusion is surprising, but not as surprising as the estimate of the costs of overlong testing.

12

# Waiting for Achilles

A famous paradox, attributed to the Greek mathematician Zeno, involves a race between the great hero Achilles and a lowly tortoise. In view of their vastly different speeds, the tortoise was granted a substantial head start. The race began, and in a short time Achilles had reached the tortoise's starting spot. But in that short time, the tortoise had moved slightly ahead. In the second stage of the race Achilles quickly covered that short distance, but the tortoise was not stationary and he moved a little further onward. And so they continued – Achilles would reach where the tortoise had been, but the tortoise would always inch ahead, just out of his reach. From this example, the great Aristotle, concluded that, "In a race, the quickest runner can never overtake the slowest, since the pursuer must first reach the point whence the pursued started, so that the slower must always hold a lead."

The lesson that we should take from this paradox is that when we focus only on the differences between groups, we too easily lose track of the big picture. Nowhere is this more obvious than in the current public discussions of the size of the gap in test scores between racial groups. In New Jersey the gap between the average scores of white and black students on the well-developed scale of the tests of the NAEP has shrunk by only about 25 percent over the past two decades. The conclusion drawn was that even though the change is in the right direction, it is far too slow.

But focusing on the difference blinds us to a remarkable success in education over the past twenty years. Although the direction and size of student improvements occur across many subject areas and many age

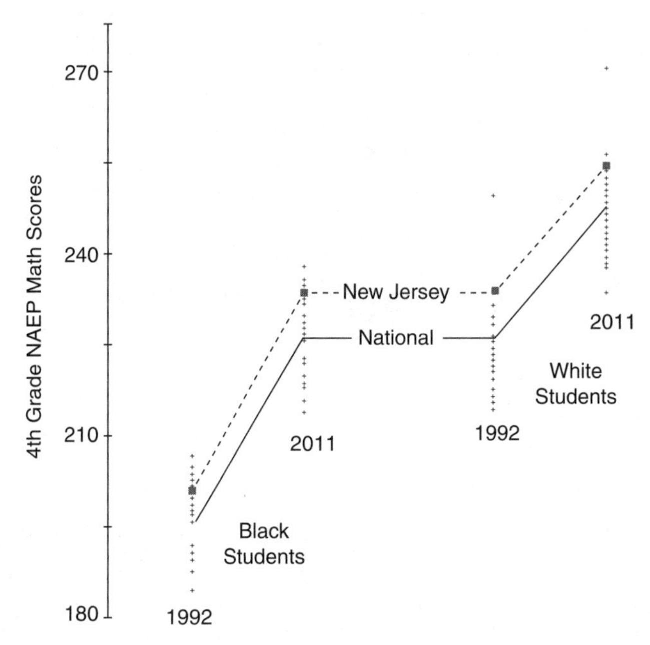


FIGURE 12.1. State average fourth grade mathematics scores of black and white students in 1992 and 2011 on the NAEP.

*Source*: U.S. Department of Education, Institute of Education Sciences, National Center for Education Statistics, NAEP, 1992, 2000, and 2011 Mathematics Assessments.

groups, I will describe just one – fourth grade mathematics. The dots in Figure 12.1 represent the average scores for all available states on NAEP's fourth grade mathematics test (with the nation as a whole as well as the state of New Jersey's dots labeled for emphasis), for black students and white students in 1992 and 2011. Both racial groups made steep gains over this time period (somewhat steeper gains for blacks than for whites). Of course, we can also see the all-too-familiar gap between the performances of black and white students, but here comes Achilles. New Jersey's black students performed as well in 2011 as New Jersey's white students did in 1992. Given the consequential difference in wealth between these two groups, which have always been inextricably connected with student performance, reaching this mark is an accomplishment worthy of applause, not criticism.

Importantly, we also see that the performance of New Jersey's students was among the very best of all states in both years and in both ethnic groups.

If we couple our concerns about American education and the remarkable success shown in these data, it seems sensible to try to understand what was going on, so that we can do more of it. A detailed examination of this question is well beyond my goals in this chapter, but let me make one observation. A little more than thirty years ago, several lawsuits were working their way through the courts that challenged the fairness of local property taxes as the principal source of public school financing. In California it was *Serrano v. Priest* and in New Jersey it was *Abbott v. Burke*; there were others elsewhere. The courts decided that, in order for the mandated "equal educational opportunity" to be upheld, per pupil expenditures in all school districts should be about equal. In order for this to happen, given the vast differences in the tax base across communities, the states had to step in and augment the school budgets of poorer districts. The fact that substantially increased funding has accompanied these robust improvements in student performance must be considered as a prime candidate in any search for cause.

This conclusion, albeit with a broader purview, was expressed by the famous economist John Kenneth Galbraith, whom I paraphrase,

> Money is not without its advantages and the case to the contrary, although it has often been made, has never proved widely persuasive.

# 13

# How Much Is Tenure Worth?

Recently, the budget crises that affect the U.S. federal, state, and local governments have led to a number of proposals that are remarkable in one way or another.[1] It seems *apropos* to examine New Jersey Governor Cristie's campaign to get rid of tenure for teachers.[2] Although there are many ways of characterizing the value of tenure, I will focus here on the possible effect of its removal on education budgets.

The fiscal goal of removing tenure is to make it easier, during periods of limited funds, for school administrators to lay off more expensive (i.e., more senior/tenured) teachers in favor of keeping less experienced/ cheaper ones. Without the protection of tenure, administrators would not have to gather the necessary evidence, which due process requires, to terminate a senior teacher. Thus, it is argued, school districts would have more flexibility to control their personnel budgets. Is this an approach likely to work?

Let us begin to examine this by first looking backward at why the policy of giving public schoolteachers tenure evolved in the first place. The canonical reason given for tenure is usually to protect academic freedom,

[1] This work was made possible through the help of Mary Ann Awad (NYSUT), David Helfman (MSEA), Rosemary Knab (NJEA), and Harris Zwerling (PSEA), who provided the data from New York, Maryland, New Jersey, and Pennsylvania, respectively. They have my gratitude: *Sine quibus non*.

[2] Governor Christie is not alone in proposing changes to long-standing progressive policies. Most notorious is Wisconsin Governor Scott Walker's attempt to eliminate collective bargaining among state employees. I assume that many, if not all, of these efforts, which seem only tangentially related to the fiscal crisis, reflected the same spirit expressed by then White House Chief-of-Staff Rahm Emanuel's observation that one ought not waste a serious crisis.

to allow teachers to provide instruction in what might be viewed as controversial topics. This is surely true, but battles akin to those that resulted from John Scopes's decision to teach evolution in the face a dogmatic school board in Dayton, Tennessee, are, happily, rare. But the reason that most teachers would want tenure is because it provides them with increased job security in general, and, in particular, as protection against capricious personnel decisions.

A more interesting question is why did states agree to grant tenure in the first place? In fact, local school boards had no direct say in the matter, for it was mandated by the state. The state officials who made this decision must have known that it would reduce flexibility in modifying school staff, and that it would make following due process in terminating a tenured teacher more time consuming and expensive. Why is tenure almost uniformly agreed to in all states? I don't know for sure, but let me offer an educated guess.[3] I am sure that most progressive officials appreciate, and value, the importance of academic freedom. But that is not the most pressing practical reason. They recognize that for teachers, tenure is a job benefit, much like health insurance, pensions, and sick time. As such it has a cash value. But it is a different kind of benefit, for unlike the others it has no direct cost. Sure, there are extra expenses when a tenured teacher is to be terminated. But if reasonable care is exercised in hiring and promotion, such expenses occur very rarely. So, I conclude, tenure was instituted to save money, exactly the opposite of what is being claimed by those who seek to abolish it.

Who is right? Is Governor Christie or am I? Happily this is a question that, once phrased carefully, is susceptible to empirical investigation. The answer has two parts. The first part is the title of this chapter: *How Much is Tenure Worth?* The second part is: Do we save enough by shifting the salary distribution of staff to compensate for the cost of its removal? I have no data on the latter and so will focus on the former.

How can we determine how much tenure is worth? One way would be to survey teachers and ask them how much money we would have to give them in order for them to relinquish it. The responses to such a

[3] As I mentioned in the introduction to this section, I served for five years as an elected member of the Princeton Board of Education; two of those years as chair of the board's Personnel Committee.

survey would surely be complex. Teachers at various stages in their careers would value it differently. Someone very close to retirement might not care very much, whereas someone in mid-career might insist on a much larger number. Teachers in subject areas, like mathematics, in which there is a shortage of qualified instructors, might not ask for much. Others, like elementary schoolteachers, in which there is a large supply, might hold tenure very dear indeed. One thing is certain: the likely answer to such a survey is going to be many thousands of dollars.

A second way to estimate the value of tenure is to take a free-market approach and run an experiment in the spirit of those described in Chapter 5. Suppose we run the following experiment: we select a few school districts, say five or six, and designate them the experimental districts, in which teachers do not have tenure. Then we pair them with an equal number of districts, matched on all of the usual variables that characterize a school district (size, ethnic mixture, safety, budget, academic performance, parental education, etc.). These we designate as the control districts, and they do have tenure. We also try to have each experimental district geographically near a matched control district.

Now we run the experiment. All of the faculty members from all the districts are put in one giant pool, and the personnel directors of each district are directed to sample from that pool of teachers in order to staff their district. This continues until the pool is empty and all of the districts are fully staffed. At this point we look at a number of dependent variables. The most obvious one is the difference in mean salary between the two groups. We might also want to look at the differences in the conditional distributions of salary, after conditioning on subject expertise and years of experience. I suspect that nontenure districts would pay more for teachers at a given level of training and experience, but be forced to concentrate their recruiting at the lowest end of the scale.

Although I am sure that the experimental (nontenure) districts will have to pay more for their staff, the question of importance is how much more. I suspect that if budgets are held constant, the nontenure districts may run out of money before they are fully staffed.

I am not sanguine that such an experiment is going to be done any time soon, although I would welcome it – perhaps so too might the teachers' unions. Of course, how could a free-market governor object?

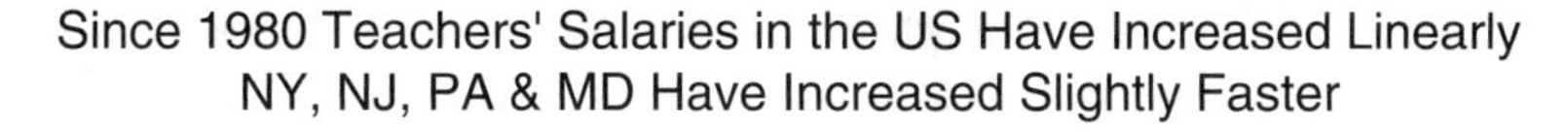


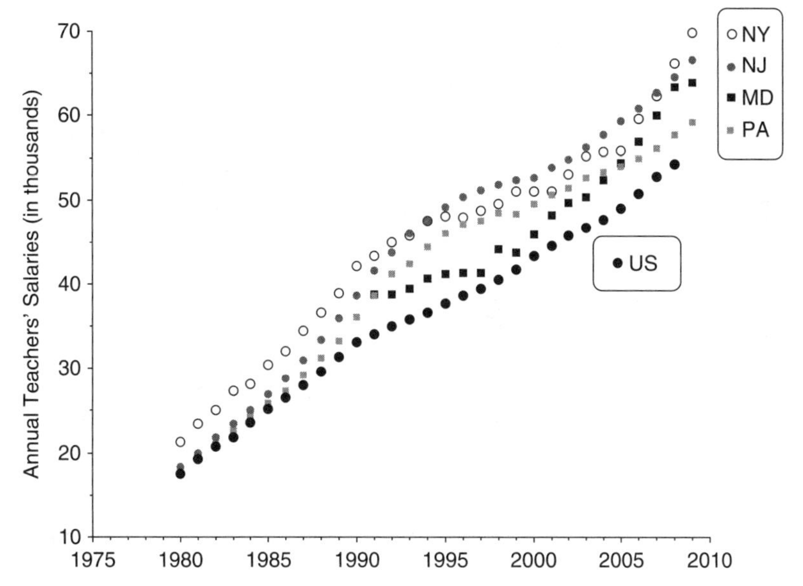


FIGURE 13.1. Mean teachers' salaries for four states and for the United States, 1980–2009.

Happily, we actually have some observational data that shed light on this subject. To get a running start on the kinds of data that can help, let us consider the teachers' salaries shown in Figure 13.1. In it are shown the annual mean salaries of public schoolteachers for four northeastern states along with that of the entire United States. The average salary increase[4] for the four states was just slightly less than $1,500/year (with New Jersey the highest of the four with a mean increase of $1,640). For the United States as a whole, it was about $1,230. The increases over these thirty years have a strong linear component.

These data can serve as a background for the rest of the discussion. Now Figure 13.2 repeats the New Jersey results for teachers and adds the mean salary for school superintendents. As we can see, superintendents make more than teachers. In 1980 they made a little bit more, but by 2009 they made a great deal more. Superintendents' salaries increased, on

[4] Actually, the average slope of the fitted regression lines for each state.

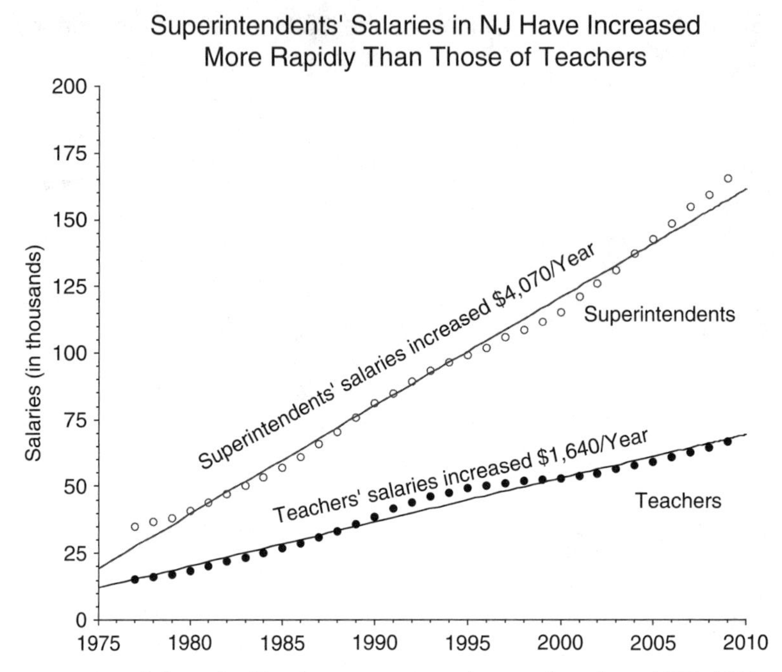


FIGURE 13.2. Salaries for New Jersey superintendents and teachers, 1977–2009.

average, more than $4,000/year. Figure 13.2 shows the increasing disparity between the salaries of superintendents, but it leaves veiled the point at which this increase began. For this, we need to look at a different plot.

In Figure 13.3 I have plotted the ratio of average superintendents' salaries to average teachers' salaries for the past thirty-three years. We quickly see that in 1977 the average superintendent in New Jersey earned about 2.25 times as much as the average teacher, but this disparity was diminishing, so that in the early 1990s the average superintendent was earning just twice what the average teacher earned. Then the disparity began to increase sharply, so that by 2009 superintendents' pay was two-and-a-half times that of teachers. What happened that should occasion this dramatic change? In 1991 the New Jersey legislature eliminated tenure for superintendents. Figure 13.3 shows a three- or four-year lag between the removal of tenure and the relative increase in superintendents' wages, which is likely due to the need for existing three- or four-year superintendent contracts to expire before they could renegotiate their salaries in the new nontenure environment. But once this happened, it is clear that

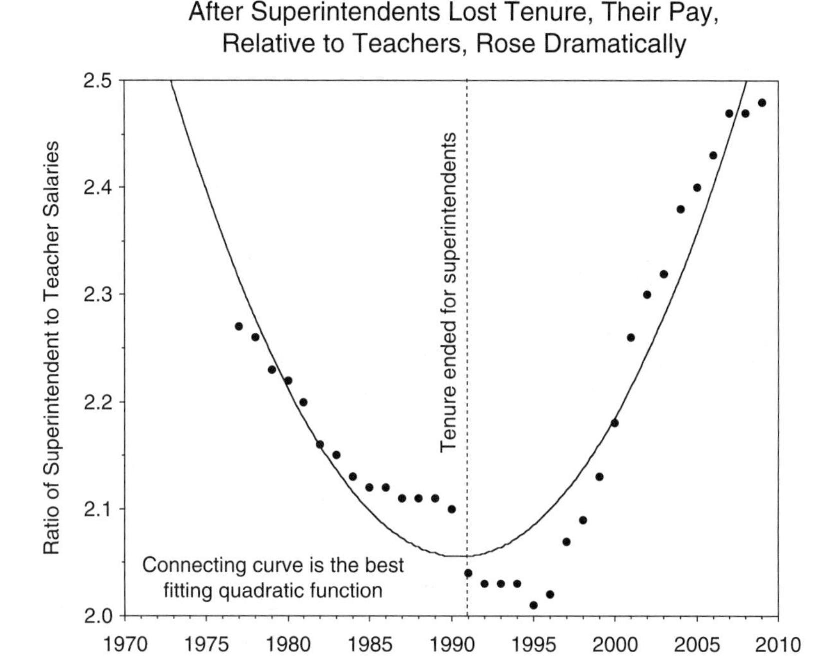


FIGURE 13.3. Salaries of school superintendents in New Jersey, relative to those of teachers, soared after career tenure was removed in 1991.

superintendents felt that they had to be compensated for the loss of tenure, and that compensation involved tens of thousands of dollars.

And now the real irony. New Jersey's legislators are well aware of the cost of eliminating tenure. On August 19, 2010 Assemblywoman Joan Voss, vice chair of the Assembly Education Committee, said in a press release that "removing lifetime tenure for superintendents has had some serious unintended consequences.... Bloated salaries and over-inflated, lavish compensation packages taxpayers are currently forced to fund." To remedy this, she and Assemblyman Ralph Caputo proposed legislation (A-2359) that would reinstitute the system of career tenure for school superintendents, which existed prior to 1991.

# 14

# Detecting Cheating Badly

## *If It Could Have Been, It Must Have Been*[1]

### Introduction

My annual retreat in the Adirondack Mountains of upstate New York was interrupted by a phone call from an old friend. It seemed that he was hired as a statistical consultant for an unfortunate young man who had been accused of cheating on a licensing exam. My friend, a fine statistician, felt that he could use a little help on the psychometric aspects of the case. After hearing the details, I agreed to participate. The situation that led to the young man's problem was directly due to fallacies associated with the unprincipled exploratory analyses of big data. The accusers had access to a huge data set from which they drew an unjustified conclusion because they did not consider the likelihood of false positives. As discussed in Chapter 5, we see again how a little thought should precede the rush to calculate.

### Fishing for Cheaters

The young man had taken, and passed, a licensing test. It was the third time he took the exam – the first two times he failed by a small amount,

[1] *Si cela aurait pu être, il doit avoir été.* (I would like to have translated this into Latin, something akin to the well-known *"Post hoc, ergo propter hoc,"* but my Latin is not up to the task, hence the French version. I would be grateful for classical help.)

but this time he passed, also by a small amount. The licensing agency, as part of their program to root out cheating, did a *pro forma* analysis in which they calculated the similarity between all pairs of examinees based on the number of incorrect item responses those examinees had in common. For the eleven thousand examinees they calculated this for all sixty million pairs.[2] After completing this analysis, they concluded that forty-six examinees (twenty-three pairs) were much more similar than one would expect by chance. Of these twenty-three pairs, only one took the exam at the same time in the same room, and they sat but one row apart. At this point a more detailed investigation was done in which their actual test booklets were examined. The test booklets were the one place where examinees could do scratch work before deciding on the correct answer. The investigator concluded on the basis of this examination that there was not enough work in the test booklets to allow him to conclude that the examinee had actually done the work, and so reached the decision that the examinee had copied and his score was not earned. His passing score was then disallowed, and he was forbidden from applying to take this exam again for ten years. The second person of the questioned pair had sat in front, and so it was decided that she could not have copied, and hence she faced no disciplinary action. The examinee who was to be punished brought suit.

## Industry Standards

There are many testing companies. Most of them are special-purpose organizations that deal with one specific exam. Typically their resources are too limited to allow the kinds of careful, rigorous research required to justify the serious consequences that investigation of cheating might lead to. The Educational Testing Service (ETS), whose annual income of well more than a billion dollars is the six hundred pound gorilla. ETS has carefully studied methods for the detection of cheating and has led the way in the specification of procedures detailed in a collaborative effort of the three major professional organizations that are concerned with

[2] They deleted about eight hundred examinees because their response patterns suggested that they guessed on a large proportion of the items and just chose the same option repeatedly, but this and other similar details are of little consequence for this narrative.

testing.[3] Most smaller testing organizations simply adopt the methods laid out by ETS and the other large organizations. To deviate from established practices would require extensive (and expensive) justification. The key standards for this situation[4] are:

1. Statistical evidence of cheating is almost never the primary motivator of an investigation. Usually there is an instigating event, for example, a proctor reporting seeing something odd, reports of examinees boasting of having cheated, other examinees reporting knowledge of collusion, or an extraordinarily large increase in score from a previous time. Only when such reports are received and documented is a statistical analysis of the sort done in this case prepared – but as confirmation. Such analyses are not done as an exploratory procedure.
2. The subject of the investigation is the score, not the examinee. After an examination in which all evidence converges to support the hypothesis that cheating might have occurred, the examinee receives a letter that states that the testing company cannot stand behind the validity of the test score and hence will not report it to whoever required the result.
3. The finding is tentative. Typically the examinee is provided with five options, taken directly from ETS (1993).
    (i) The test taker may provide information that might explain the questioned circumstances (e.g., after an unusually large jump in score the test taker might bring a physician's note confirming severe illness when the test was originally taken). If accepted the score obtained is immediately reported.
    (ii) The test taker may elect to take an equivalent retest privately, at no cost, and at a convenient time to confirm the score being reviewed. If the retest score is within a specified range of the questioned score, the original score is confirmed. If it is not the test taker is offered a choice of other options.

[3] The American Educational Research Association, the American Psychological Association, and the National Council on Measurement in Education jointly publish *Standards for Educational and Psychological Testing*. The most recent edition appeared in 1999. The committee who prepared each edition of the *Standards* was largely made up of scholars who at one time or another had spent significant time at one (or more) of the major testing organizations.

[4] The best public source for both the operational details of a cheating investigation and its ethical and legal motivation is a 1993 report prepared for the ETS Board of Trustees. The preceding synthesis was derived from that report.

(iii) The test taker may choose independent arbitration of the issue by the American Arbitration Association. The testing organization pays the arbitration fee and agrees to be bound by the arbitrator's decision.
(iv) The test taker may opt to have the score reported accompanied by the testing organization's reason for questioning it and the test taker's explanation.
(v) The test taker may request cancellation of the score. The test fee is returned and the test taker may attend any regularly scheduled future administration.

Option (ii) is, by far, the one most commonly chosen.

## Why Are These Standards Important?

Obviously, all three of these standards were violated in this summer's investigation. And, I will conclude, the testing organization ought to both change its methods and revise the outcome of the current case. To support my conclusion, let me illustrate, as least partially, why time, effort, and the wisdom borne of experience have led serious testing organizations to scrupulously follow these guides. The example, using mammograms for the early detection of breast cancer among women, illustrates how the problem of false positives is the culprit that bedevils attempts to find rare events.

## False Positives and Mammograms

Annually about 180,000 new cases of invasive breast cancer are diagnosed in women in the United States. About forty thousand of these women are expected to die from breast cancer. Breast cancer is second only to skin cancer as the most commonly diagnosed cancer, and second only to lung cancer in death rates. Among U.S. women, about one in four cancers are breast cancer, and one out of every eight U.S. women can expect to be diagnosed with breast cancer at some time in their lives.

However, some progress in the battle against the horrors of breast cancer has been made. Death rates have been declining over the past twenty years through a combination of early detection and improved treatment.

The first steps in early detection are self-examination and mammograms. The strategy is then to investigate any unusual lumps found by these relatively benign procedures with more invasive, yet revealing methods – most particularly a biopsy.

How effective are mammograms? One measure of their effectiveness is characterized in a simple statistic. If a mammogram is found to be positive, what is the probability that it is cancer? We can estimate this probability from a fraction that has two parts. The numerator is the number of breast cancers found, and the denominator is the number of positive mammograms. The denominator contains both the true and the false positives.

The numerator first: it is 180,000 cases.

The denominator has two parts: the true positives, 180,000, plus the false positives. How many of these are there? Each year thirty-seven million mammograms are given in the United States. The accuracy of mammograms varies from 80 percent to 90 percent depending on circumstances.[5] For this discussion let's assume the more accurate figure of 90 percent accuracy. This means that when there is a cancer, 90 percent of the time it will be found, and when there is no cancer, 90 percent of the time it will indicate no cancer. But this means that 10 percent of the time it will indicate a possible cancer when there is none. So, 10 percent of thirty-seven million mammograms yields 3.7 million false positives. And so the denominator of our fraction is 180,000 plus 3.7 million, or roughly 3.9 million positive mammograms.

Therefore the probability of someone with a positive mammogram having breast cancer is 180,000/3.9 million or about 5 percent. That means that 95 percent of the women who receive the horrible news that the mammogram has revealed something suspicious, and that they must return for a biopsy, are just fine.

[5] There is a huge research literature on breast cancer, much of which looks into this very question. But the results vary. One recent study (Berg et al. 2008) found that mammography had an accuracy of 78 percent, but when combined with ultrasound was boosted to 91 percent. So the figure that I use of 90 percent accuracy for mammograms alone does no damage to the reputation of mammography. Of key importance, this 90 percent figure is the probability of finding a cancer given that it is there. But this is not what we want. We are interested in the inverse question, what is the probability of cancer given that the test says it is there. To do this formally requires a simple application of Bayes's Theorem, which I do, informally, in the subsequent derivation.

Is a test with this level of accuracy worth doing? The answer to this question depends on many things, but primarily it depends on the costs of doing mammograms versus the costs of not doing them.[6]

The standards used in doing mammograms follow industry standards of testing. Obviously the 5 percent accuracy would be much worse if the denominator was larger. We get a strong hint that the first standard is followed when we note that it is never recommended that the entire population of the United States have annual mammograms. Instead there is some prescreening based on other characteristics. Modern standards suggest that only women with a family history of breast cancer or are more than fifty years of age get mammograms, and consideration is now being given to revising these to make them more restrictive still.

The third standard is also followed (with option (ii) the most frequently chosen one even in this situation), in that anyone tagged as possibly having a tumor is retested more carefully (usually with another mammogram and then a biopsy using thin-needle aspiration).

Suppose mammogram policy followed the methods suffered by the unfortunate victim of the current investigation. First, instead of thirty-seven million mammograms with 3.7 million false positives, we would have three hundred million people tested yielding thirty million false positives. This would mean that about 99.5 percent of all positive mammograms would be wrong – making the test almost worthless as a diagnostic tool.

Suppose after a positive mammogram, instead of continued testing, we moved straight to treatment including, but not limited to, a mastectomy with radiation and chemotherapy. Obviously, this would be a remarkably dopey idea.

Yet, how much smarter is it to effectively ban someone from a field for which they have been training for several years? This is especially unwise, when retesting is a cheap and easy alternative.

[6] The cost associated with not doing a mammogram is measured in the number of women who would die unnecessarily. In the past, survival was dependent on early detection, thus justifying the widespread use of mammograms. But in 2010, a large Norwegian study that examined the efficacy of modern, directed therapy showed that there was essentially no difference in survival rates of women who had mammograms and those who did not (Kalager et al. 2010). This led to the suggestion that mammograms use be much more limited among asymptomatic women.

## How Accurate Is the Cheater Detection Program in Question?

The short answer is, "I don't know." To calculate this we would need to know (1) how many cheaters there are – analogous to how many cancer tumors, and (2) how accurate is the cheating detection methodology. Neither of these numbers is known; they are both missing. Let us treat them as we would any missing data and, through multiple imputations, develop a range of plausible values. I will impute a single value for these unknowns but invite readers to stick in as many others as you wish to span the space of plausibility.

Let us assume that there are one hundred cheaters out of the eleven thousand examinees. If this number seems unreasonable to you, substitute in another value that you find more credible. And, let us assume that the detection scheme is, like a well done mammogram, 90 percent accurate (although because the reliability of the test is about 0.80, this is a remarkably Pollyannaish assumption). Let us further assume that the errors are symmetric – it is as accurate at picking out a cheater as it is at identifying an honest examinee. This too is unrealistic because the method of detection has almost no power at detecting someone who only copied a very few items, or who, by choosing a partner wisely, only copied correct answers. Nevertheless, we must start someplace; so let's see where these assumptions carry us.

We are trying to estimate the probability of being a cheater given that the detection method singles us out as one.

The numerator of this probability is ninety; the one hundred honest-to-goodness cheaters are identified with 90 percent accuracy. The denominator is the true cheaters identified – 90, plus the 1,100 false positives (10 percent of the eleven thousand honest examinees). So the fraction is 90/1,190 = 7.6 percent.

Or, given these assumptions, more than 92 percent of those identified as cheaters were falsely accused! If you believe that these assumptions are wrong, change them and see where it leads. The only assumption that must be adhered to is that no detection scheme is 100 percent accurate. As Damon Runyon pointed out, "nothing in life is more than 3 to 1."

In the current situation the pool of positive results was limited to forty-six individuals. How can we adapt this result to the model illustrated by the mammogram example? One approach might be to treat those forty-six as being drawn at random from the population of the 1,190 yielded by our example, thus maintaining the finding that only 7.6 percent of those identified as cheaters were correctly so identified.

But a different interpretation might be to believe that the testing organization recognized that their detection methodology had little power in finding an examinee that only copied a few answers from a neighbor. This reality might have suggested that the cost of missing such a cheater was minor. Because all they could detect were those who had not only copied a substantial number of items, but also were unfortunate enough to have copied a substantial number of wrong answers. With this limited goal in mind they set a dividing line for cheating so that only forty-six people out of eleven thousand were snared as possible. Then they were forced to conclude that twenty-two of the twenty-three pairs of examinees were false positives because of the physical impossibility of copying. Thus, despite the uncontroversial evidence that at least twenty-two of twenty-three were false positives, they concluded that if it could be, then it was. It is certainly plausible that all twenty-three pairs were false positives and that one of them happened to occur in the same testing center. This conclusion is bolstered by the facts that the validity of only one section of the test was being questioned and that the test taker passed the other sections.

The scanning of the test booklet for corroborating evidence is a red herring; for although we know that the one candidate in question had what were deemed insufficient supporting calculations, we don't know what the test booklets of other examinees looked like – they were all destroyed.

It seems clear from these analyses that the evidence is insufficient for actions as draconian as those imposed.

Of course, it is important for any testing organization to impose standards of proper behavior upon its examinees. A certified test score loses value if it can be obtained illegally. And yet any detection scheme is imperfect. And the costs of its imperfections are borne by the innocents accused. With due deliberation we have, in the past, determined that the inconvenience and expense involved for the vast majority of women who have an unnecessary mammogram is offset by the value of early detection for the small percentage of women whose lives are lengthened because of the early detection. But new rules, based upon improved treatment, are aimed at reducing the false positives without increasing the risk for those women with cancer.

The arguments of those opposed to capital punishment have been strengthened enormously by the more than two hundred innocent men released from long prison terms in the twenty years since DNA evidence became more widely used. This result dramatically makes the point that the imperfections of our detection methods argue against draconian measures when alternatives are available.

## Coda

After extensive and expensive legal wrangling, the testing organization grudgingly agreed to allow the candidate to retake the exam under tight security. He passed.

# 15

# When Nothing Is Not Zero

## *A True Saga of Missing Data, Adequate Yearly Progress, and a Memphis Charter School*

One of the most vexing problems in all of school evaluation is missing data. This challenge manifests itself in many ways, but when a school, a teacher, or a student is being evaluated on the basis of their performance, the most common missing data are test scores.

If we want to measure growth we need both a pre- and a postscore. What are we to do when one, the other, or both are missing? If we want to measure school performance, what do we do when some of the student test scores are missing?

This problem has no single answer. Sometimes the character of the missingness can be ignored, but usually this approach is reasonable only when the proportion of missing data is very small. Otherwise the most common strategy is "data imputation" (discussed in Chapters 3 and 4, and its misuse was shown in Chapter 7). Data imputation involves deriving some plausible numbers and inserting them in the holes. How we choose those numbers depends on the situation and on what ancillary information is available.

Let us consider the specific situation of Promise Academy, an inner-city charter school in Memphis, Tennessee, that enrolled students in kindergarten through fourth grade for the 2010–11 school year. Its performance has been evaluated on many criteria, but of relevance here is the performance of Promise students on the state's reading/language arts (RLA) test. This score is dubbed its Reading/Language Arts Adequate Yearly Progress (AYP) and it depends on two components: the scores of third- and fourth-grade students on the RLA portion of the test and the performance of fifth-grade students on a separate writing assessment.

We observe the RLA scores, but because Promise Academy does not have any fifth-grade students, all of the writing scores are missing. What scores should we impute to allow us to calculate a plausible total score? The state of Tennessee's rules require a score of zero be inserted. In some circumstances imputing scores of zero might be reasonable. For example, if a school only tests half of its students, we might reasonably infer that it was trying to game the system by choosing to omit the scores from what are likely to be the lowest-performing students. This strategy is mooted by imputing a zero for each missing score. But this is not the situation for Promise Academy. Here the missing scores are structurally missing – they could not submit fifth-grade scores because they have no fifth graders! Yet following the rules requires imputing a zero score and would mean revoking Promise Academy's charter.

What scores should we impute for the missing ones so that we can more sensibly compute the required AYP summary? Figure 15.1 shows the performance of all Tennessee schools on the third- and fourth-grade RLA tests. We see that Promise Academy did reasonably well on both the third- and fourth-grade tests.

Not surprisingly, the calculated figure for AYP can be estimated from third- and fourth-grade RLA performance alone. It will not be perfect, because AYP also includes fifth-grade performance on the writing test, but it turns out to estimate AYP very well indeed. The scatter plot in Figure 15.2 shows the bivariate distribution of all the schools with their AYP RLA score on the horizontal axis and the predicted AYP RLA score on the vertical axis. Drawn in on this plot is the regression line that provides the prediction of AYP RLA from the best linear combination of third- and fourth-grade test scores. In this plot we have included Promise Academy. As can be seen from the scatter plot, the agreement between predicted AYP and actual AYP is

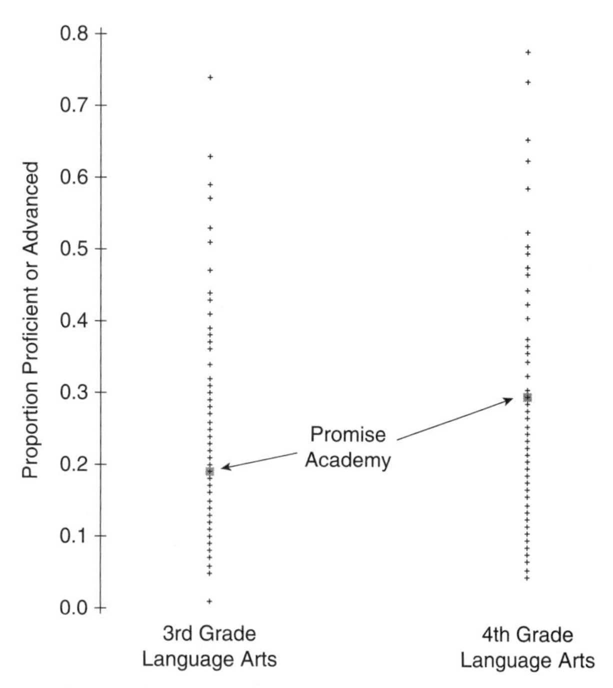


FIGURE 15.1. The distributions of all Tennessee schools on RLA for third and fourth grades with Promise Academy's scores emphasized.

remarkably strong. The correlation between the two is 0.96 and the prediction equation is:

$$AYP = .12 + 0.48 \times 3rd\ RLA + .44 \times 4th\ RLA$$

Promise Academy's predicted AYP score is much higher than its actual AYP score (0.33 predicted vs. 0.23 actual) because it does not have any fifth graders and hence the writing scores were missing. The difference between the actual and the predicted is due to imputing a zero writing score for that missing value. The fitted straight line gives the best estimate of what Promise's AYP would have been in the counterfactual case of their having had fifth graders to test. We thus estimate that they would have finished thirty-seventh among all schools in AYP rather than eighty-eighth based on the imputation of a zero.

This dramatic shift in Promise Academy's rank is clearer in Figure 15.3, which shows the two AYP scores graphically.

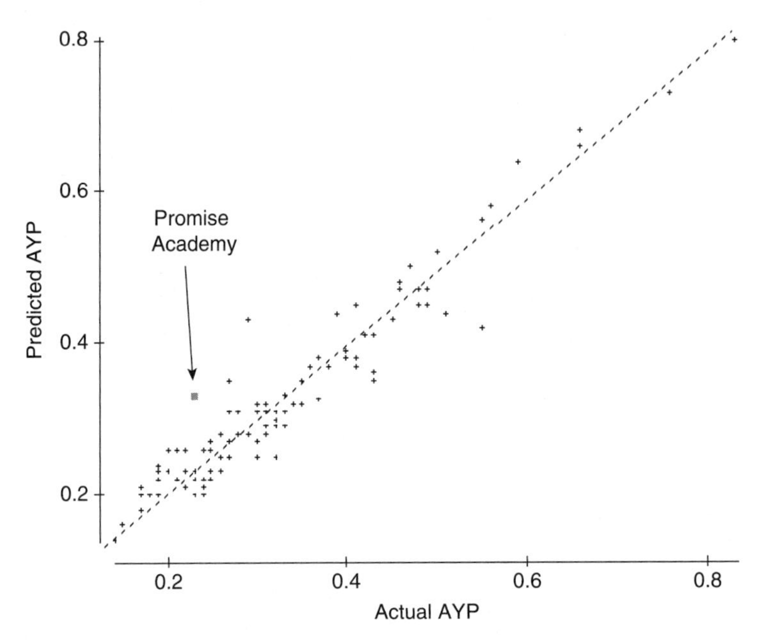


FIGURE 15.2. A scatter plot comparing the actual AYP on the horizontal axis with its predicted value obtained without using fifth-grade writing on the vertical axis.

Standard statistical practice would strongly support imputing an estimated score for the impossible-to-obtain fifth-grade writing score, rather than the grossly implausible value of zero.

The solid performance of Promise Academy's third and fourth graders in the language arts tests they took was paralleled by their performance in mathematics. In Figure 15.4 we can compare the math performance of Promise's students with those from the 110 other schools directly. This adds further support to the estimated AYP RLA score that we derived for Promise Academy.

The analyses performed thus far focus solely on the actual scores of the students at Promise Academy. They ignore entirely one important aspect of the school's performance that has been a primary focus for the evaluation of Tennessee schools for more than a decade – value added. It has long been felt that having a single bar over which all schools must pass was unfair to schools whose students pose a greater challenge for instruction. Thus, evaluators have begun to assess not just what level the students reach, but also how far they have come. To do this, a complex statistical model, pioneered by William Sanders and his colleagues, is fit to "before

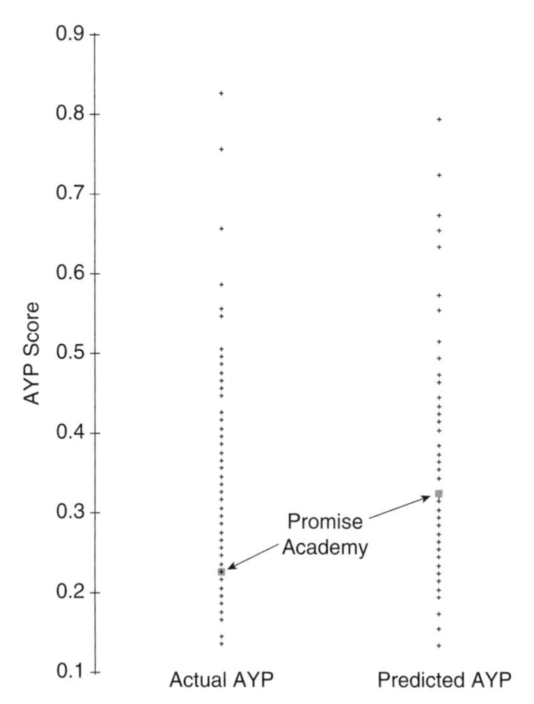


FIGURE 15.3. A comparison of the actual AYP scores based on all three years of data with its predicted value based on just third- and fourth-grade data. The difference between the two values for Promise Academy shows the effect of the zero imputation for their missing fifth grade.

and after" data to estimate the gains – the value added – that characterize the students at a particular school. That work provides ancillary information to augment the argument made so far about the performance of Promise Academy. Specifically, two separate sets of analyses by two different research organizations (Stanford's Center for Research on Education Outcomes and Tennessee's own Value-Added Assessment System) both report that in terms of value added, Promise Academy's instructional program has yielded results that place it among the best-performing schools in the state. Though many issues regarding the utility of the value-added models remain unresolved, none of them would suggest that a large gain implies a bad result.

When data are missing, there will always be greater uncertainty in the estimation of any summary statistic that has, as one component, a missing piece. When this occurs, standards of statistical practice require

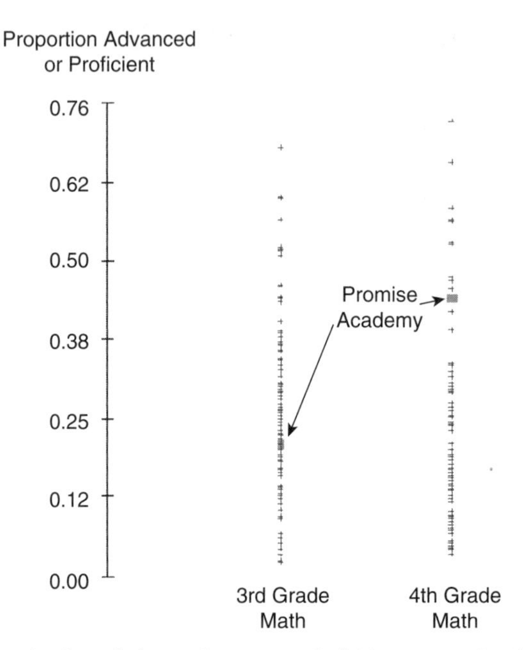


FIGURE 15.4. A display of the performance of all Tennessee schools in third- and fourth-grade mathematics shows that the performance of Promise Academy's students in RLA was not an isolated instance.

that we use all available ancillary data to estimate what was not measured. In the current case, the missing piece was the performance of fifth graders in a school that ends at fourth grade. In this instance, it is clear from third- and fourth-grade information in both RLA and mathematics, as well as from value-added school parameters, that the Promise Academy AYP score obtained by imputing a zero score for the missing fifth-grade writing assessment grossly misestimates the school's performance.

In October 2011 the school presented its case to the Memphis City Schools administration. Using all the information available, the administration recognized the competent job being done by Promise Academy. Its charter was renewed for another year, at which time Promise Academy's student body will include fifth grade, and thus the need for imputing structurally missing data will vanish.

But the lesson from this experience is important and should be remembered: *nothing is not always zero.*

# 16

# Musing about Changes in the SAT

## *Is the College Board Getting Rid of the Bulldog?*

During the first week of March 2014 the front page of most U.S. newspapers reported a story released by the College Board of big changes planned for the SAT.[1] These were amplified and augmented by the cover story of the March 9 issue of the *New York Times' Sunday Magazine* (Balf 2014).

Having spent twenty-one years as an employee of the Educational Testing Service, the College Board's principal contractor in the production, administration, and scoring of the SAT, I read these reports with great interest. In the end I was left wondering why these changes, in particular, were being proposed and, moreover, why all the hoopla.

Let me elaborate. Three principal changes described were:

1. The scoring would no longer have a penalty for guessing.
2. They would reduce the amount of arcane vocabulary used on the test.
3. They would discontinue the "writing" part of the exam and return to just the canonical verbal and quantitative sections, leaving writing as a stand-alone that would be optional.

By my reckoning the first change is likely to have only small effect, but it probably will increase the error variance of the test, making the scores a little less accurate. The second change addresses something that isn't really

[1] My gratitude to longtime ETS Research Scientist Brent Bridgeman for his guidance through the details of the modern SAT, although he is not responsible for where I have taken his wisdom.

much of a problem, but it is always wise to be vigilant about including item characteristics that are unrelated to the trait being measured. I am not sanguine about any changes being implemented successfully. The third change is probably a face-saving reaction to the 2005 introduction of the writing section, which did not work as hoped, and is the one modification that is likely to yield a positive effect.

To see how I arrived at these conclusions, let's go over each of the changes in more detail.

## No Penalty for Guessing

The current SAT uses what is called "formula scoring." That is, an examinee's score is equal to the number right diminished by one-fourth the number wrong (for the kinds of five-choice items commonly used on the SAT). The idea behind this is that if examinees guess completely at random among the choices on the items for which they don't know the answer, for every four items[2] they get wrong they will, on average, get one right by chance. So, under these circumstances, the expected gain from guessing is zero. Thus there is neither any benefit to guessing nor does guessing add bias to the scores, although there is the binomial variance due to guessing that is unnecessarily added to the score. Note that if an examinee has partial knowledge, and can eliminate one or more of the distractors (the wrong answers), the expected gain from guessing, even after correction, is positive, thus giving some credit for such partial knowledge. The proposed change would do away with this correction and simply use the unadjusted number right as the input for the scoring algorithm.

What is likely to be the effect? My memory is that the correlation between formula score and number correct score is very close to one. So whatever change occurs will probably be small. But, maybe not, for it seems plausible that formula scoring deters at least some random guessing. Such guessing adds no information to the score, just noise, so it is hard to make a coherent argument for why we would want to encourage it. But perhaps it was decided that if the effect of making the change is

[2] Or (k-1) for a k-choice items.

small, why not do it – perhaps it would make the College Board look responsive to critics without making any real change.

## Reduce the Amount of Arcane Vocabulary

This modification has been the subject of considerable discussion (see Murphy's December 2013 *Atlantic* article), but the meaning of arcane, in this context, remains shrouded in mystery. Let us begin with a dictionary definition.

Definition: arcane (adjective) – known or understood by very few; mysterious; secret; obscure; esoteric:

> She knew a lot about Sanskrit grammar and other arcane matters.

Arcane words, defined in this way, bring to mind obscure words used in very narrow contexts, such as *chukka* or *cwm*. The first is one of the seven-and-a-half-minute periods that make up a polo match and is rumored to have last been used as part of an SAT more than sixty years ago; the second derives from the Welsh word for *valley*, and its principal use is in the closing plays of Scrabble games.

But this does not seem to characterize what critics of SAT vocabulary have in mind. An even dozen words that have been used to illustrate this "flaw" are, in alphabetical order: artifice, baroque, concomitant, demagogue, despotism, illiberal, meretricious, obsequious, recondite, specious, transform, and unscrupulous.

Words of this character are less often heard in common conversation than in reading. Thus, I think a better term for such words is not *arcane* but rather *literary*. Wanting to rid the SAT of the lexical richness of words accumulated through broad reading seems hard to justify. I will not try. Instead let me parse the topic into what I see as its three component parts.

(1) How much arcane vocabulary is on the SAT? I suspect that, using the true definition of *arcane*, there is close to none. Using my modified definition of "literary" vocabulary, there is likely some; but with the promised shift to including more "foundational" documents on the test (e.g., Declaration of Independence, Federalist Papers), it seems unavoidable that certain kinds of literary, if not arcane, vocabulary

will show up. In an introductory paragraph of Alexander Hamilton's *Federalist #1* I found a fair number of my illustrative dozen (indicated in **boldface** in the below box).

*Federalist #1 General Introduction* by Alexander Hamilton

"An **over-scrupulous** jealousy of danger to the rights of the people, which is more commonly the fault of the head than of the heart, will be represented as mere pretense and **artifice**, the stale bait for popularity at the expense of the public good. It will be forgotten, on the one hand, that jealousy is the usual **concomitant** of love, and that the noble enthusiasm of liberty is apt to be infected with a spirit of narrow and **illiberal** distrust. On the other hand, it will be equally forgotten that the vigor of government is essential to the security of liberty; that, in the contemplation of a sound and well-informed judgment, their interest can never be separated; and that a dangerous ambition more often lurks behind the **specious** mask of zeal for the rights of the people than under the forbidden appearance of zeal for the firmness and efficiency of government. History will teach us that the former has been found a much more certain road to the introduction of **despotism** than the latter, and that of those men who have overturned the liberties of republics, the greatest number have begun their career by paying an **obsequious** court to the people; commencing **demagogues**, and ending tyrants."

(2) Is supporting the enrichment of language with unusual words necessarily a bad thing? I found Hamilton's "General Introduction" to be lucid and well argued. Was it so in spite of his vocabulary? Or because of it? James Murphy, in his December 2013 *Atlantic* article, "The Case for SAT Words," argues persuasively in support of enrichment. I tend to agree.

(3) But some pointlessly arcane words may still appear on the SAT and rarely appear anywhere else (akin to such words as *busker*, which has appeared in my writing for the first time today). If such vocabulary

is actually on the SAT, how did it find its way there? This question probably has many answers, but I believe that they all share a common root. Consider a typical item on the verbal section of the SAT (or any other verbal exam) – say a verbal reasoning item or a verbal analogy item. The difficulty of the item varies with the subtlety of the reasoning or the complexity of the analogy. It is a sad, but inexorable fact about test construction that item writers cannot write items that are more difficult than they are smart. And so the distribution of item difficulties looks a great deal like the distribution of item-writer ability. But, in order to discriminate among candidates at high levels of ability, test specifications require a fair number of difficult items. How is the item writer going to respond when her supervisor tells her to write ten hard items? Often the only way to generate such items is to dig into a thesaurus and insert words that are outside broad usage. This strategy will yield the result that fewer people will get them right (the very definition of *harder*).

Clearly the inclusion of such vocabulary is not directly related to the trait being tested (e.g., verbal reasoning), any more than making a writing task harder by insisting that examinees hold the pen between their toes. And so, getting rid of such vocabulary may be a good idea. I applaud it. But how, then, will difficult verbal items be generated? One possibility is to hire much smarter item writers; but such people are not easy to find, nor are they cheap. The College Board's plan may work for a while, as long as unemployment among Ivy League English and classics majors is high. But as the job market improves, such talent will become rarer. Thus, so long as the need for difficult verbal items remains, I fear we may see the inexorable seepage of a few arcane words back onto the test. But with all the checks and edits a prospective SAT item must negotiate, I don't expect to see many.

## Making the Writing Portion Optional

To discuss this topic fully, we need to review the purposes of a test. There are at least three:

1. Test as contest – the higher score wins (gets admitted, gets the job, etc.). For this purpose to be served the only characteristic a test must have is fairness.

2. Test as measuring instrument – the outcome of the test is used for further treatment (determination of placement in courses, measurement of the success of instruction, etc.). For this purpose the test score must be accurate enough for the applications envisioned.
3. Test as prod – Why are you studying? I have a test. Or, more particularly, why does the instructor insist that students write essays? Because they will need to write on the test. For this purpose the test doesn't even have to be scored, although that practice would not be sustainable.

With these purposes in mind, why was the writing portion added to the core SAT in 2005? I don't know. I suspect for a combination of reasons, but principally (3), as a prod.[3] Why a prod and not for other purposes? Scoring essays, because of its inherent subjectivity, is a very difficult task on which to obtain much uniformity of opinion. More than a century ago it was found that there was as much variability among scorers of a single essay as there was across all the essays seen by a single scorer. After uncovering this disturbing result the examiners reached the conclusion that scorers needed to be better trained. Almost twenty-five years ago a more modern analysis of a California writing test found that the variance component for raters was the same as that for examinees. So a hundred years of experience training essay raters didn't help.

A study done in the mid-1990s used a test made up of three thirty-minute sections. Two sections were essays, and one was a multiple-choice test of verbal ability. The multiple-choice score correlated more highly with each of the essay scores than the essay scores did with each other. This means that if you want to predict how an examinee will do on some future writing task, you can do so more accurately with a multiple-choice test than a writing test.

[3] This conclusion is supported in the words of Richard Atkinson, who was president of the University of California system at the time of the addition of the writing section (and who was seen as the prime mover behind the changes). He was quite explicit that he wanted to use it as a prod – "an important aspect of admissions tests was to convey to students, as well as their teachers and parents, the importance of learning to write and the necessity of mastering at least 8th through 10th grade mathematics …" and "From my viewpoint, the most important reason for changing the SAT is to send a clear message to K–12 students, their teachers and parents that learning to write and mastering a solid background in mathematics is of critical importance. The changes that are being made in the SAT go a long way toward accomplishing that goal." http://www.rca.ucsd.edu/speeches/CollegeAdmissionsandtheSAT-APersonalPerspective1.pdf (accessed August 24, 2015).

Thus, I conclude that the writing section must have been included primarily as a prod, so that teachers would emphasize writing as part of their curriculum. Of course, the writing section also included a multiple-choice portion (which was allocated thirty-five of the sixty minutes for the whole section) to boost the reliability of the scores up to something approaching acceptable levels.

The writing section has also been the target of a fair amount of criticism from teachers of writing, who claimed, credibly, at least to me, that allocating twenty-five minutes yielded no real measure of a student's ability. This view was supported by the sorts of canned general essays that coaching schools had their students memorize. Such essays contained the key elements of a high-scoring essay (four hundred words long, three quotations from famous people, seven complex words, and the suitable insertion of some of the words in the "prompt" that instigated the essay).

Which brings us to the point where we can examine what has changed since the writing section was instituted to cause the College Board to reverse field and start removing it. I suspect that at least part of the reason was that the College Board had unreasonable expectations for a writing task that could be administered in the one hour available for it. It was thus an unrealistic task that was expensive to administer and score, and it yielded an unreliable measure of little value.

Making it optional, as well as scoring it on a different scale than the rest of the SAT, is perhaps the College Board's way of getting rid of it gracefully and gradually. Considering the resources planned for the continued development and scoring of this section, it appears that the College Board is guessing that very few colleges will require it and few students will elect to take it.

## Coda

The SAT has been in existence, in one form or another, since 1926. Its character was not arrived at by whim. Strong evidence, accumulated over those nine decades, supports many of the decisions made in its construction. But it is not carved in stone, and changes have occurred continually. However, those changes were small, inserted with the hope of making an improvement if they work and not being too disastrous if they do not.

This follows the best advice of experts in quality control and has served the College Board well. The current changes fall within these same limits. They are likely to make only a very small difference, but with luck the difference will be a positive one. The most likely place for an improvement is the shrinkage of the writing section. The other two changes appear to be largely cosmetic and not likely to have any profound effect. Why were they included in the mix?

Some insight into this question is provided by recalling a conversation in the late 1960s between John Kemeny and Kingman Brewster, the presidents of Dartmouth and Yale, respectively. Dartmouth had just finalized the plans to go coed and successfully avoided the ire of those alumni who inevitably tend to oppose any changes. Yale was about to undergo the same change, and so Brewster asked Kemeny whether he had any advice. Kemeny replied, "Get rid of the bulldog."

At the same time that Dartmouth made the enrollment change, they also switched their mascot from the Dartmouth Indian to the Big Green. Alumni apparently were so up in arms about the change in mascot that they hardly noticed the girls. By the time they did, it was a *fait accompli* (and they then noticed that they could now send their daughters to Dartmouth and were content).

Could it be that the College Board's announced changes vis-à-vis guessing and arcane vocabulary were merely their bulldog, which they planned to use to divert attention from the reversal of opinion represented by the diminution of importance of the writing section? Judging from the reaction in the media to the College Board's announcement, this seems a plausible conclusion.

17

# For Want of a Nail

## *Why Worthless Subscores May Be Seriously Impeding the Progress of Western Civilization*

*How's your wife?*
*Compared to what?*
Henny Youngman

Standardized tests, whether to evaluate student performance in coursework, to choose among applicants for college admission, or to license candidates for various professions, are often marathons. Tests designed to evaluate knowledge of coursework typically use the canonical hour, admissions tests are usually two to three hours, and licensing exams can take days. Why are they as long as they are?[1] The first answer that jumps immediately to mind is the inexorable relationship between a test's length and its reliability.[2] And so to know how long a test should be we must

[1] Issues of overtesting have been prominently in the news since the implementation of "No Child Left Behind" because it is felt that too much of students' instructional time was being used for testing. The argument I make here is certainly applicable to those concerns but is more general. I will argue that for most purposes tests, and hence testing time, can be reduced substantially and still serve those purposes.

[2] Reliability is a measure of the stability of a score, essentially telling you how much the score would change were the person to take another test on the same subject later, with all else kept constant. In Chapter 5 I defined reliability in a heuristic way, as the amount of evidence brought into play to support a claim. High reliability typically implies a great deal of evidence. This definition fits well here, where the longer the test, the more reliable, *ceteris paribus*.

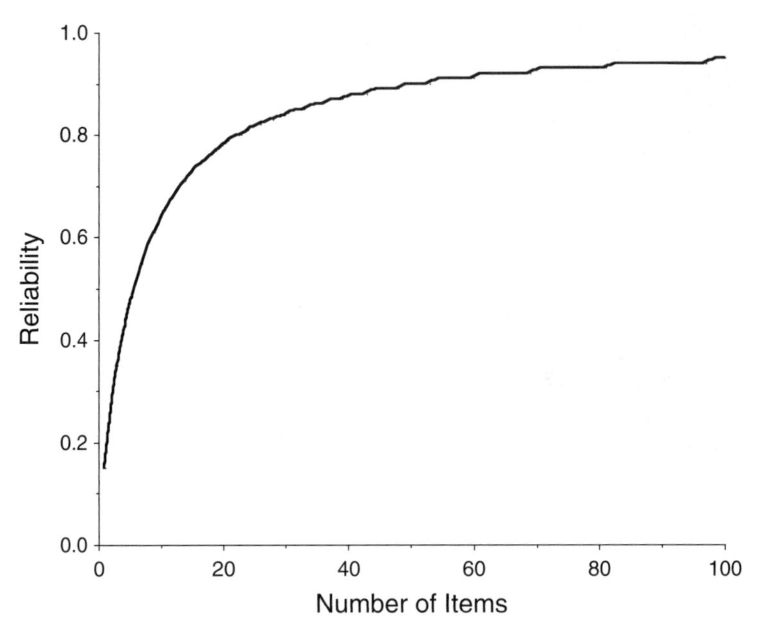


FIGURE 17.1. Spearman-Brown function showing the reliability of a test as a function of its length, if a one-item test has a reliability of 0.15.

first ask, "How much reliability is needed?" Which brings us to Henny Youngman's quip, compared to what? There is a long answer to this, but let us begin with a short one, "a shorter test."

Although a test score always gets more reliable as the test generating it gets longer, *ceteris paribus*, the law of diminishing returns sets in very quickly and so the marginal gain in reliability with increases in test length gets small in a big hurry. In Figure 17.1 we show the reliability of a typical professionally prepared test as a function of its length. It shows that the marginal gain of moving from a thirty-item test to a sixty- or even ninety-item one is not worth the trouble unless such additional reliability is required. As we will show shortly, it is hard to think of a situation where that would be the case, so why are tests as long as they are? What is to be gained through the excessive length? What is lost?

## A Clarifying Example – the U.S. Census

Our intuitions can be clarified with an example, the Decennial U.S. Census. On midnight of January 1, 2010, the year of the last Census, it was estimated that there were 308,745,538 souls within the borders of the United States. The error bounds were ± 31,000. The budget of the 2010 Census was $13 billion, or approximately $42 per person. Is it worth this amount of money to just get a single number? Before answering this question, note that a single clerk, with access to an adding machine in a minute or two, could estimate the change from the previous Census (to an accuracy of ±0.1 percent).[3]

It doesn't take a congressional study group and the Office of Management and the Budget to tell us "no." If all the Census gave us was just that single number it would be a colossal waste of taxpayer money. However, the constitutionally mandated purpose of the Census is far broader than just providing a single number. It must also provide the estimates used by states to allocate congressional representation, as well as much, much narrower estimates (small area estimates, like "how many households with two parents and three or more children live in the Bushwick section of Brooklyn?"). In statistics these are called *small area estimates*, and they are crucial for the allocation of social services and for all sorts of other purposes. The Census provides such small area estimates admirably well, and because it can do so, makes it worth its impressive cost.

## Back to Tests

Now let us return to tests. Instead of characterizing cost in terms of dollars (a worthwhile metric, for sure, but grist for another mill) let us use instead examinee time. Is it worth using an hour (or two or more) of examinee time to estimate just a single number – a single score? Is the small marginal increase in accuracy obtained from a sixty- or eighty-item test over, say a thirty-item test, worth the extra examinee time?

A glance at the gradual slope of the Spearman-Brown curve shown in Figure 17.1 as it nears its upper bound tells us that we aren't getting much of a return on our investment. And multiplying the extra hour spent by

[3] This remarkable estimation comes from periodic surveys that tell us that the net increase in the U.S. population is one person every thirteen seconds, and so to get an accurate estimate of the total population at any moment one merely needs to ascertain how much time has elapsed since the last estimate, in seconds, divide by thirteen, and add in that increment.

each examinee by the millions of examinees that often take such tests makes this conclusion stronger still. What would be the circumstances in which a test score with a reliability of 0.89 will not suffice, but one of .91 would? Off hand, it is hard to think of any, but we will return to this in greater detail later.

Perhaps there are other uses for the information gathered by the test that require additional length; the equivalent of the small area estimates of Census. In testing such estimates are usually called *subscores*, which are really small area estimates on various aspects of the subject matter of the test. On a high school math test these might be subscores on algebra, arithmetic, geometry, and trigonometry. For a licensing exam in veterinary medicine there might be subscores on the pulmonary system, the skeletal system, the renal system, and so on. There is even the possibility of cross-classified subscores in which the same item is used on more than one subscore – perhaps one on dogs, another on cats, and others on cows, horses, and pigs. Such cross-classified subscores are akin to the Census having estimates by ethnic group and also by geographic location.

The production of meaningful subscores would be a justification for tests that contain more items than would be required merely for an accurate enough estimate of total score. What is a meaningful subscore? It is one that is reliable enough for its prospective use and one that has information that is not adequately contained in the total test score.

There are at least two prospective uses of such subscores:

(1) To aid examinees in assessing their strengths and weaknesses, often with an eye toward remediating the latter, and
(2) To aid individuals and institutions (e.g., teachers and schools) in assessing the effectiveness of their instruction, again with an eye toward remediating weaknesses.

In the first case, helping examinees, the subscores need to be reliable enough so that attempts to remediate weaknesses do not become just the futile pursuit of noise. And, obviously, the subscore must contain information that is not available from the total score. Let us designate these two characteristics of a worthwhile subscore: *reliability* and *orthogonality.* Subscores' reliability is governed by the same inexorable rules of reliability as overall scores – as the number of items they are based on decreases, so too does their reliability. Thus if we need reliable subscores we must have enough items for that purpose. This would mean that the overall test's length would have to be greater than would be necessary for merely a single score.

For the second use, helping institutions, the test's length would not have to increase, for the reliability would be calculated over the number of individuals from that institution who took the items of interest. If that number was large enough the estimate could achieve high reliability.

And so it would seem that one key justification for what appears at first to be the excessive lengths of most common tests is to provide feedback to examinees in subscores calculated from subsets of the tests. How successful are test developers in providing such subscores?

Not particularly, for such scores are typically based on few items and hence are not very reliable. This result led to the development of empirical Bayes methods to increase the reliability of subscores by borrowing strength from other items on the test that empirically yield an increase in reliability.[4] This methodology often increased the reliability of subscores substantially, but at the same time the influence of items from the rest of the test reduced the orthogonality of those subscores to the rest of the test. Empirical Bayes gaveth, but it also tooketh away. What was needed was a way to measure value of an augmented subscore that weighed the delicate balance between increased reliability and decreased orthogonality.

Until such a measure became available the instigating question "how successful are test developers in providing useful subscores?" would still remain unanswered.

Happily, the ability to answer this important question was improved markedly in 2008 with the publication of Shelby Haberman's powerful new statistic that combined both reliability and orthogonality.[5] Using this tool Sandip Sinharay[6] searched high and low for subscores that had added value over total score, but came up empty. Sinharay's empirical results were validated in simulations he did that matched the structure

4 This methodology was proposed by Wainer, Sheehan, and Wang in 2000, and was later elaborated in chapter 9 of Thissen and Wainer's 2001 book *Test Scoring*.

5 Haberman 2008.

6 Sinharay 2010.

commonly encountered in different kinds of testing situations. Sinharay's results were enriched and expanded by Richard Feinberg,[7] and again the paucity of subscores worth having was confirmed. This same finding, of subscores adding no marginal value over total score, was reconfirmed for tests whose goal was classification.[8] While it is too early to say that there are no subscores that are ever reported that are worth having, it seems sensible that unless tests are massively redesigned such subscores are likely rare.

Surprisingly, at least to me, the search for subscores of value to institutions also seems to have been futile[9] largely because most institutions that were attempting to use such scores had fewer than fifty examinees, and so those scores too were neither reliable nor orthogonal enough to lift their usefulness above the bar set by total score.

## If Not for Subscores, Is There Another Justification for Long Tests?

Where does this leave us? Unless we can find a viable purpose for which unreliable and nonorthogonal subscores have marginal value over just the total test score, we are being wasteful (and perhaps unethical) to continue to administer tests that take more examinee time than is justified by the information yielded.

As we saw in Chapter 16, one possible purpose is the use of the test as a prod to motivate students to study all aspects of the curriculum and for the teachers to teach it. Surely, if the test is much shorter, fewer aspects of the curriculum will be well represented. But this is easily circumvented in a number of ways. If the curriculum is sampled cleverly neither the teachers nor the examinees will know exactly what will be on the test and so have to include all of it in their study. Another approach is to follow NAEP's lead and use some sort of clever design in which all sectors of the curriculum are well covered but no single examinee will get all parts. That

[7] Feinberg 2012.
[8] Sinharay 2014.
[9] Haberman, Sinharay, and Puhan 2009.

will allow estimates of subarea mastery to be estimated in the aggregate and, through the statistical magic of score equating, still allow all examinee scores to rest on a common metric. We should keep in mind the result that has been shown repeatedly with adaptive tests in which we can give a test of half its usual length with no loss of motivation. Sure there are fewer items of each sort, but examinees must still study all aspects because on a shorter test each item "counts" more toward the final score.

So, unless evidence can be gathered that shows a radical change in teaching and studying behavior with shorter tests, we believe that we can reject motivation as a reason for giving too long tests.

Another justification for the apparent excessive lengths of tests is that the small additional reliability due to that extra length is of practical importance. Of course, the legitimacy of such a claim would need to be examined on a case-by-case basis, but perhaps we can gain some insight through a careful study of one artificial situation that has similarities to a number of serious tests.

Let us consider the characteristics of a prototypical licensing examination that has, say, three hundred items, takes eight hours to administer, and has a reliability of 0.95. Such characteristics show a marked similarity to a number of professional licensing exams. Let's suppose it is being used to license attorneys (it might just as well be physicians or veterinarians or nurses or certified public accountants). The purpose of such an exam is to make a pass-fail decision and let us assume that the passing score is 63 percent correct.

To make this demonstration dramatic let's see what happens to the accuracy of our decisions if we make truly draconian reductions in the test length. To begin with let's eliminate 75 percent of the test items and reduce the test to just seventy-five items. Because of the gradual slope of the reliability curve shown in Figure 17.1, this kind of reduction would only shrink the reliability to 0.83. Is this still high enough to safeguard the interests of future clients? The metric of reliability is not one that is close to our intuitions, so let us shift to something easier to understand – how many wrong pass-fail decisions would be made.

With the original test 3.13 percent of the decisions would be incorrect and these would be divided between false positives (passing when

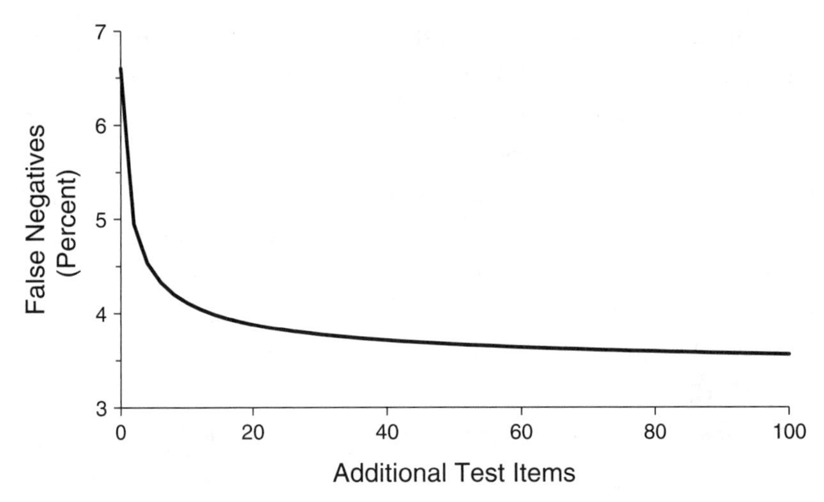


FIGURE 17.2. The improvement in the false negative rate yielded through the lengthening of the test for those who only marginally failed.

they should have failed) of 1.18 percent and false negatives (failing when they should have passed) of 1.95 percent. How well does our shrunken test do? First, the overall error rate increases to 6.06 percent, almost double what the longer test yielded. This breaks down to a false positive rate of 2.26 percent and a false negative rate of 3.80 percent. Is the inevitable diminution of accuracy sufficiently large to justify the fourfold increase in test length? Of course, that is a value judgment, but before making it we must realize that the cost in accuracy can be eased. The false positive rate for this test is the most important one, for it measures the proportion of incompetent attorneys that incorrectly are licensed. Happily, we can control the false positive rate by simply raising the passing score. If we increase the passing score to 65 percent instead of 63 percent the false positive rate drops back to the same 1.18 percent we had with the full test. Of course, by doing this, the false negative rate grows to 6.6 percent, but this is a venial sin that can be ameliorated easily by adding additional items to those candidates who only barely failed. Note that the same law of diminishing returns that worked to our advantage in total score (shown in Figure 17.1) also holds in determining the marginal value of adding more items to decrease the false negative rate. The parallel to the Spearman-Brown curve is shown in Figure 17.2.

TABLE 17.1. *Summary of Passing Statistics*

| Test Length | Passing Score (percent) | Reliability | Total Error Rate | False Positive Rate | False Negative Rate |
|---|---|---|---|---|---|
| 300 | 63 | 0.95 | 3.13 | 1.18 | 1.95 |
| 75 | 63 | 0.83 | 6.06 | 2.26 | 3.80 |
| 75 | 65 | 0.83 | 7.78 | 1.18 | 6.60 |
| 40 | 63 | 0.72 | 8.23 | 2.59 | 5.64 |
| 40 | 66 | 0.72 | 11.62 | 1.14 | 10.48 |

The function in Figure 17.2 shows us that by adding but forty items for only those examinees just below the cut score (those whose scores range from just below the minimal passing score of 65 percent to about 62 percent) we can attain false negative rates that are acceptably close to those obtained with the three-hundred-item test. This extension can be accomplished seamlessly if a computer administers the test. Thus for most of the examinees their test would take but one-fourth the time as it would have previously, and even for the small number of examinees who had to answer extra items, these were few enough so that even they still came out far ahead.

In Table 17.1 we show a summary of these results as well as parallel results for an even more dramatically reduced test form with but forty items.

Thus, at least for simple pass-fail decisions, it seems that we can dismiss accuracy as a reason for the excessive test lengths used; for within plausible limits we can obtain comparable error rates with much shorter tests, albeit with an adaptive stopping rule.

As it turns out this is also true if a purpose of the test score is to pick winners. This is a more complex process than simply seeing if any score is above or below some fixed passing score, for now we must look at a pair of scores and decide which candidate is better. A detailed examination of this aspect is beyond the goals of this chapter, but I will discuss it in detail in a subsequent account.

## The Costs of Using Excessively Long Tests and the Progress of Western Civilization

Costs can be measured in various ways and, of course, they accrue differentially to different portions of the interested populations: the testing organization, the users of the scores, and the examinees.

The cost to the users of the scores is nil because neither their time nor money is used to gather the scores.

The cost to the testing organization is likely substantial because a single operational item's cost is considerably greater than $2,500. Add to this the costs of "seat time" paid to whoever administers the exam, grading costs, and so forth, and it adds to a considerable sum. Fixed costs being what they are, giving a new test of one-fourth the length of an older one does not mean one-fourth the cost, but it does portend worthwhile savings. We are also well aware that concerns of test speededness at current lengths could be ameliorated easily if the time allowed for the test was shrunken, but not quite as far as the number of items would suggest.

Which brings us to the examinees. Their costs are of two kinds: (1) the actual costs paid to the testing organization, which could be reduced if the costs to that organization were dramatically reduced, and (2) the opportunity costs of time.[10]

If the current test takes eight hours then a shortened form of only one-fourth its length might, *ceteris paribus*, be only two hours long, a savings of six hours per examinee. Multiplied by perhaps one hundred thousand examinees who annually seek legal licensure, this would yield a time savings of six hundred thousand hours. Keeping in mind that the examinees taking a bar exam are (or shortly will be) all attorneys. What can be accomplished with six hundred thousand extra hours of attorneys' time?

Think of how much good six hundred thousand annual hours of *pro bono* legal aid could do – or a like amount of effort from fledging accountants at tax time, architects, professional engineers, physicians, or veterinarians. It is not an exaggeration to suggest that this amount of spare time from pre- or just-licensed professionals could accelerate the progress of our civilization.

[10] These opportunity costs, if translated to K–12 schoolchildren, should be measured in lost instructional time.

## Conclusions

The facts presented here leave us with two possibilities:

(1) To shorten our tests to the minimal length required to yield acceptable accuracy for the total score, and thence choose more profitable activities for the time freed up, or
(2) To reengineer our tests so that the subscores that we calculate have the properties that we require.

> The fault, dear Brutus, is not in our statistical measures, but in our tests, that they are poorly designed.
>
> *Cassius to Brutus in Shakespeare's Julius Caesar (Act 1, Scene II)*

It has been shown[11] clearly that the shortcomings found in the subscores calculated on so many of our large-scale tests are due to flaws in the tests' designs. Thus the second option, the one we find most attractive, requires redesigning our tests. A redesign is necessary because, in a parallel to what Cassius so clearly pointed out to Brutus more than four centuries ago, we cannot retrieve information from our tests if the capacity to gather that information was not built in to begin with. Happily, a blueprint for how to do this was carefully laid out in 2003 when Bob Mislevy, Linda Steinberg, and Russell Almond provided the principles and procedures for what they dubbed "Evidence Centered Design." It seems worth a shot to try it. In the meantime we ought to stop wasting resources giving tests that are longer than the information they yield is worth.

Finally, we must not forget a key lesson illustrated by the Census: while reliable small area estimates can be made, they are not likely to come cheap.

## One Caveat

It may be that such reengineering will not help if the examinee population is too homogeneous. Consider a health inventory. Surely everyone would agree that recording both height and weight is important. While the two variables are correlated, each contains important information not

[11] Sinhary, Haberman, and Wainer 2011.

found in the other; a person who weighs 220 pounds is treated differently if he is 6'5" than if he is 5'6". Additionally both can be recorded with great accuracy. So, by the rules of valuable subscores, both should be kept in the inventory. But suppose the examinees were all men from a regiment of marines. Their heights and weights would be so highly correlated that if you know one you don't need the other. This may be the case for such highly selected populations as attorneys or physicians and so there is likely little prospect for worthwhile subscores. But there remains the strong possibility of advancing civilization.

SECTION IV

# Conclusion: ~~Don't~~ Try This at Home

For most of its history, science shared many characteristics of a sect. It had its own language, strict requirements for membership, and its own closely held secrets. Science can also usefully be thought of as a metaphorical mountain, existing at an altitude far above where ordinary people lived. Those who practiced it shared the common asceticism of the scientific method, while communicating with one another in the too often mysterious language of mathematics.

This description of science has been accurate for at least the 2,500 years since the time of the Pythagoreans but is now in the midst of a revolutionary change. This shift is occurring because it has become apparent to everyone that the methods of science have to be more widely used for humans to understand the world we inhabit.

One of the principal purposes of this book has been to illustrate how often it is that we can use scientific methods without the burden of complex mathematics or arcane methodologies. It is like learning how to ride a bicycle without knowing the equations of motion that describe how to balance upright on just two wheels. Stanford's Sam Savage has an evocative phrase, only slightly inaccurate, that for some tasks we can learn better through the seat of our pants than we can through the seat of our intellect. Obviously, both parts of our anatomy are often useful to master a broad range of situations.

Most of the knotty problems discussed so far were unraveled using little more than three of the essential parts of scientific investigations:

(1) Some carefully gathered data, combined with
(2) Clear thinking and

(3) Graphical displays that permit the results of the first two steps to be made visible.

Of course, many of the problems encountered by scientists are not susceptible to amateurs – coronary bypass surgery, the design of nuclear reactors, and genetic decoding of the Ebola virus are three that come immediately to mind; there are many others. This is not surprising. What is remarkable is how broad the range of problems is that are susceptible to illumination by thoughtful and industrious amateurs.

Let me offer three real examples:

## 1. Why "Market Forces" Have Not Controlled the Cost of Health Care

It has often been argued that the reason that free-market ideas of competition have failed to deliver in the health care industry is because the costs of medical care are not available to patients in advance of the procedure. This was illustrated by Jaime Rosenthal, Xin Lu, and Peter Cram who, in 2013, tried, unsuccessfully, to find out what would be the cost of a hip replacement. Their findings were replicated and amplified by *New York Times* science reporter Gina Kolata who, on behalf of her pregnant and uninsured daughter, called a number of hospitals to try to find out how much the delivery would cost. She was not particularly successful until she identified herself as a *Times* reporter and got through to the Kennedy Health System's president and chief executive, Martin A. Bieber, who told her that a normal delivery is approximately $4,900, and newborn care costs about $1,400. And, he added, Kennedy charges all uninsured patients those prices, which are 115 percent of the Medicare rates, no matter what their income. She also reported that Dartmouth Medical Center, one of the few places that posts its prices, charges the uninsured about $11,000 for a normal delivery and newborn care.

However, one can argue that such procedures as hip replacements or even giving birth are complex and idiosyncratic, thus the costs, which depend heavily on the specific situation, might vary enormously. This variation may make cost estimates too difficult to predict accurately. In addition, it is uncertain what strictures about revealing prices to the public are placed on hospital employees.

This issue was discussed around the dinner table at the Bernstein household in Haverford, Pennsylvania, recently. Jill Bernstein (age 14), advised and guided by her father Joseph (an orthopedic surgeon) devised an investigation that she felt would shed additional light on the situation.

Ms. Bernstein called twenty local (Philadelphia) hospitals and explained she needed an electrocardiogram (ECG) but had no insurance. She then asked how much would one cost. An ECG is simple and its costs ought not to vary by case. Seventeen of the twenty hospitals refused to tell her and the three that did quoted prices of $137, $600, and $1,200, respectively (even surpassing the vast proportional variation in delivery costs that Gina Kolata found). Then (and here's Ms. Bernstein's genius) she called the hospitals back and said that she was coming in for an ECG and asked how much did parking cost. Nineteen of the twenty provided parking costs (ten offered free or discounted parking, indicating an awareness of consumer's concerns about cost). She then wrote up and published her results[1] and was subsequently interviewed on National Public Radio.

## 2. How Often Is an Active Cell Phone in a Car Like an Open Beer Can?

In 2003 a study by Utah psychologist David Strayer and his colleagues found that "people are as impaired when they drive and talk on a cell phone as they are when they drive intoxicated at the legal blood-alcohol limit" of 0.08 percent.[2] This result was derived from a simulated driving task in which the subjects of the experiment had to "drive" while either talking on the phone or after having ingested some alcohol.

This finding also sparked discussion around the Bernstein dinner table. Now it was Jill's sixteen-year-old brother James who was interested in further investigation. He was curious how often drivers had their cell

1 Bernstein and Bernstein 2014.
2 This is the minimum level that defines illegal drunken driving in most U.S. states.

phones active and whether there was any relationship between having the phone at the ready and such other variables as:

(1) Having the seat belt fastened,
(2) Driver's gender, and
(3) The type of vehicle.

To gather data he set up a chair at a local intersection and recorded how many drivers, while waiting at the red light for about thirty seconds, used the time to speak or text on their phones. He found about 20 percent (of the thousand cars he recorded) did and that amount was unrelated to sex of driver or type of vehicle, but it was related to seatbelt use (the relationship between the two was in the expected direction). A fair proportion continued speaking as they drove off. He concluded that having an active cell phone in a car should be regarded in much the same way as an "open bottle" of alcohol – a dangerous temptation.[3]

## 3. How Do Monarch Butterflies Come North Each Spring?

In the early 1970s Fred Urquhart and his colleagues at the University of Toronto used a tiny tag to identify Monarch Butterflies (*Danaus plexippus*) and so to track them to their winter home in Mexico.[4] With the assistance of Kenneth Brugger, a textile engineer working in Mexico, they found their destination. It was a site in the Sierra Madre mountains west of Mexico City that boasted a huge number of milkweed trees completely covered by orange Monarchs enjoying the warm weather until it was time for their return north in the spring.

The final step that would complete the story of the Monarch's migration was to track all of the butterflies on the trip north. But this step was beyond the resources of this research team – they were flummoxed.

Thus far this butterfly study is typical of many scientific investigations – being run by experts using fancy techniques and high-tech

[3] Bernstein & Bernstein, 2015.
[4] My thanks to Lee Wilkinson (2005) for this example and for some of the description.

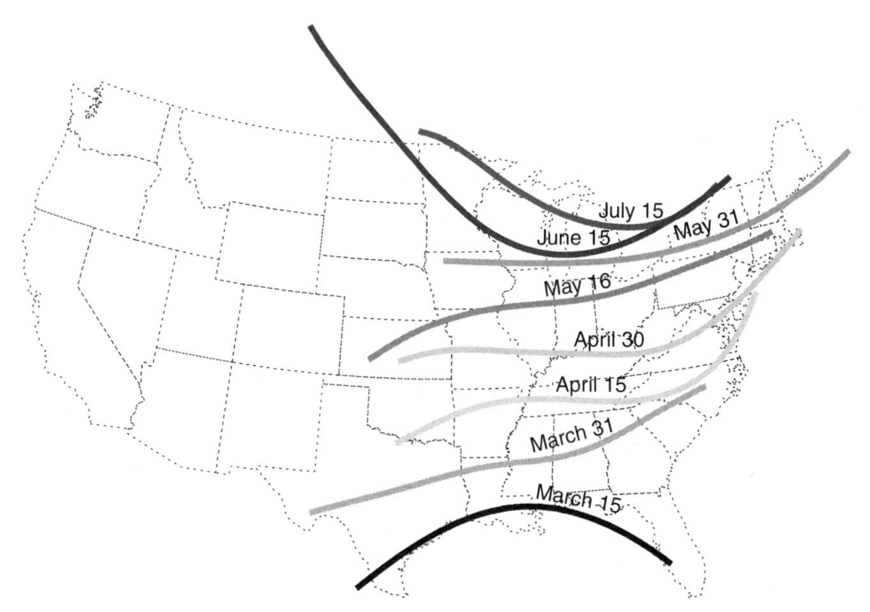


FIGURE C.1. Migration pattern of Monarch butterflies returning north after wintering in Mexico. See color version facing page 109.

instrumentation. But all of this horsepower was insufficient for the final step, for to accomplish it would require following all of the swarms of butterflies when they left Mexico and keeping track of their many different pathways – both where they went and when they got there; a task well beyond the resources available.

Yet in 1977 appeared a summary display (see Figure C.1) of exactly where and when Monarchs traveled on their way north. How was this accomplished?

> I have always depended on the kindness of strangers.
>
> *Blanche Dubois*

It will come as no surprise, given the theme of this section, that the cavalry that rode to the rescue of this investigation were hundreds of children, teachers, and other observers that reported to the Journey North Project under the auspices of the Annenberg/CPB Learner Online program. Each data point in the graph is the location and date of an observer's first sighting of an adult Monarch during the first six months of 1977.

## Coda

These three examples show how the methods of science have descended from the ivory tower and can be used by anyone with sufficient grit and wit to contribute to their personal understanding of the world and, if the results are important enough, to the general understanding.

I have reached the end of my tale. Each chapter was meant to convey a little of the way a modern data scientist thinks, and how to begin to solve what may have seemed like impossibly vexing problems. Underlying all of this are some deep ideas about the critical role in science played by evidence and some of its important characteristics. Especially crucial are (1) explicit hypotheses, (2) evidence that can test those hypotheses, (3) reproducibility, and (4) the control of hubris.

Let me conclude with the words of three scientists and a pitcher who emphasized these characteristics in memorable ways.

The pitcher first – Paraphrasing the great Satchel Page's comment on the value of money:

> Evidence may not buy happiness,
> but it sure does steady the nerves.

This idea was augmented and emphasized by the chemist August Wilhelm von Hofman (1818–92), the discoverer of formaldehyde:

> I will listen to any hypothesis, but on one condition –
> that you show me a method by which it can be tested.

A century later one key aspect of scientific evidence, reproducibility, was made explicit by the journal editor G. H. Scherr (1983):

> The glorious endeavour that we know today as science has grown out of the murk of sorcery, religious ritual, and cooking. But while witches, priests, and chefs were developing taller and taller hats, scientists worked out a method for determining the validity of their results; they learned to ask Are they reproducible?

And finally, on the crucial importance of modesty and controlling the natural hubris that would have us draw inferences beyond our data, from the beloved Princeton chemist Hubert N. Alyea:

> I say not that it is, but that it seems to be;
> As it now seems to me to seem to be.

Professor Alyea's observation seems like a remarkably apt way for me to express my own feelings as I end this book.

# Bibliography

American Educational Research Association, the American Psychological Association, and the National Council on Measurement in Education (1999). *Standards for Educational and Psychological Testing.* Washington, DC: American Psychological Association.

Amoros, J. (2014). Recapturing Laplace. *Significance* 11(3): 38–9.

Andrews, D. F. (1972). Plots of high-dimensional data. *Biometrics* 28: 125–36.

Angoff, C., and Mencken, H. L. (1931). The worst American state. *American Mercury* 31: 1–16, 175–88, 355–71.

Arbuthnot, J. (1710). An argument for divine providence taken from the constant regularity in the births of both sexes. *Philosophical Transactions of the Royal Society* 27: 186–90.

Austen, J. (1817). *North to the Hanger, Abbey.* London: John Murray, Albermarle Street.

Balf, T. (March 9, 2014). The SAT is hated by – all of the above. *New York Times Sunday Magazine*, 26–31, 48–9.

Beardsley, N. (February 23, 2005). The Kalenjins: A look at why they are so good at long-distance running. *Human Adaptability.*

Berg, W. A., et al. (2008). Combined screening with ultrasound and mammography vs. mammography alone in women at elevated risk of breast cancer. *Journal of the American Medical Association* 299(18): 2151–63.

Bernstein, J. R. H., and Bernstein, J. (2014). Availability of consumer prices from Philadelphia area hospitals for common services: Electrocardiograms vs. parking. *JAMA Internal Medicine* 174(2): 292–3. http://archinte.jamanetwork.com/ (accessed December 2, 2014).

Bernstein, J. J., and Bernstein, J. (2015). Texting at the light and other forms of device distraction behind the wheel. *BMC Public Health*, 15: 968. DOI 10.1186/s12889-015-2343-8 (accessed September 29, 2015).

Berry, S. (2002). One modern man or 15 Tarzans? *Chance* 15(2): 49–53.

Bertin, J. (1973). *Semiologie Graphique.* The Hague: Mouton-Gautier. 2nd ed. (English translation done by William Berg and Howard Wainer and published as *Semiology of Graphics*, Madison: University of Wisconsin Press, 1983.)

Bock, R. D. (June 17, 1991). *The California Assessment. A talk given at the Educational Testing Service*, Princeton, NJ.

Bridgeman, B., Cline, F., and Hessinger, J. (2004). Effect of extra time on verbal and quantitative GRE scores. *Applied Measurement in Education* 17: 25–37.

Bridgeman, B., Trapani, C., and Curley, E. (2004). Impact of fewer questions per section on SAT I scores. *Journal of Educational Measurement* 41: 291–310.

Briggs, D. C. (2001). The effect of admissions test preparation: Evidence from NELS:88. *Chance* 14(1): 10–18.

Bynum, B. H., Hoffman, R. G., and Swain, M. S. (2013). A statistical investigation of the effects of computer disruptions on student and school scores. Final report prepared for Minnesota Department of Education, Human Resources Research Organization.

Chernoff, H. (1973). The use of faces to represent points in k-dimensional space graphically. *Journal of the American Statistical Association* 68: 361–8.

Cizek, G. J., and Bunch, M. B. (2007). *Standard Setting: A Guide to Establishing and Evaluating Performance Standards on Tests.* Thousand Oaks, CA: Sage.

Clauser, B. E., Mee, J., Baldwin, S. G., Margolis, M. J., and Dillon, G. F. (2009). Judges' use of examinee performance data in an Angoff standard-setting exercise for a medical licensing examination: An experimental study. *Journal of Educational Measurement* 46(4): 390–407.

Cleveland, W. S. (2001). Data science: An action plan for expanding the technical areas of the field of statistics. *International Statistical Review* 69: 21–6.

Cobb, L. A., Thomas, G. I., Dillard, D. H., Merendino, K. A., and Bruce, R. A. (1959). An evaluation of internal-mammary-artery ligation by a double-blind technic. *New England Journal of Medicine* 260(22): 1115–18.

Cook, R., and Wainer, H. (2016). Joseph Fletcher, thematic maps, slavery and the worst places to live in the UK and the US. In C. Kostelnick and M. Kimball (Eds.), *Visible Numbers, the History of Statistical Graphics.* Farnham, UK: Ashgate Publishing (forthcoming).

(2012). A century and a half of moral statistics in the United Kingdom: Variations on Joseph Fletcher's thematic maps. *Significance* 9(3): 31–6.

Davis, M. R., (2013, May 7). Online testing suffers setbacks in multiple states. *Education Week.* Retrieved on 30 August, 2013 from http://www.edweek.org/ew/articles/2013/05/03/30testing.h32.html.

DerSimonian, R., and Laird, N. (1983). Evaluating the effect of coaching on SAT scores: A meta-analysis. *Harvard Education Review* 53: 1–15.

Dorling, D. (2005). *Human Geography of the UK.* London: Sage Publications.

Dorling, D., and Thomas, B. (2011). *Bankrupt Britain: An Atlas of Social Change.* Bristol, UK: Policy Press.

Educational Testing Service (1993). *Test Security: Assuring Fairness for All.* Princeton, NJ: Educational Testing Service.

Feinberg, R. A. (2012). A simulation study of the situations in which reporting subscores can add value to licensure examinations. PhD diss., University of Delaware. Accessed October 31, 2012, from ProQuest Digital Dissertations database (Publication No. 3526412).

Fernandez, M. (October 13, 2012). El Paso Schools confront scandal of students who "disappeared" at test time. *New York Times.*

Fisher, R. A. (1925). *Statistical Methods for Research Workers. Edinburgh:* Oliver and Boyd.

Fletcher, J. (1849a). Moral and educational statistics of England and Wales. *Journal of the Statistical Society of London* 12: 151–76, 189–335.

(1849b). *Summary of the Moral Statistics of England and Wales.* Privately printed and distributed.

(1847). Moral and educational statistics of England and Wales. *Journal of the Statistical Society of London* 10: 193–221.

Fox, P., and Hender, J. (2014). The science of data science. *Big Data* 2(2): 68–70.

Freedle, R. O. (2003). Correcting the SAT's ethnic and social-class bias: A method for reestimating SAT scores. *Harvard Educational Review* 73(1): 1–43.

Friendly, M., and Denis, D. (2005). The early origins and development of the scatterplot. *Journal of the History of the Behavioral Sciences* 41(2): 103–30.

Friendly, M., and Wainer, H. (2004). Nobody's perfect. *Chance* 17(2): 48–51.

Galchen, R. (April 13, 2015). Letter from Oklahoma, Weather Underground: The arrival of man-made earthquakes. *The New Yorker*, 34–40.

Gelman, A. (2008). *Red State, Blue State, Rich State, Poor State: Why Americans Vote the Way They Do.* Princeton, NJ: Princeton University Press.

Gilman, R., and Huebner, E. S. (2006). Characteristics of adolescents who report very high life satisfaction. *Journal of Youth and Adolescence* 35(3): 311–19.

Graunt, J. (1662). *Natural and Political Observations on the Bills of Mortality.* London: John Martyn and James Allestry.

Haberman, S. (2008). When can subscores have value? *Journal of Educational and Behavioral Statistics* 33(2): 204–29.

Haberman, S. J., Sinharay, S., and Puhan, G. (2009). Reporting subscores for institutions. *British Journal of Mathematical and Statistical Psychology* 62: 79–95.

Halley, E. (1686). An historical account of the trade winds, and monsoons, observable in the seas between and near the tropicks; with an attempt to assign the physical cause of said winds. *Philosophical Transactions* 183: 153–68. The issue was published in 1688.

Hand, E. (July 4, 2014). Injection wells blamed in Oklahoma earthquakes. *Science* 345(6192): 13–14.

Harness, H. D. (1837). *Atlas to Accompany the Second Report of the Railway Commissioners, Ireland.* Dublin: Irish Railway Commission.

Hartigan, J. A. (1975). *Clustering Algorithms*. New York: Wiley.

Haynes, R. (1961). *The Hidden Springs: An Enquiry into Extra-Sensory Perception*. London: Hollis and Carter. Rev. ed. Boston: Little, Brown, 1973.

Hill, R. (2013). An analysis of the impact of interruptions on the 2013 administration of the Indiana Statewide Testing for Educational Progress – Plus (ISTEP+). http://www.nciea.org/publication_pdfsRH072713.pdf.

Hobbes, T. (1651). *Leviathan, or the matter, forme, and power of a commonwealth, ecclesiasticall and civill*. Republished in 2010, ed. Ian Shapiro (New Haven, CT: Yale University Press).

Holland, P. W. (October 26, 1980). *Personal communication*. Princeton, NJ.

(1986). Statistics and causal inference. *Journal of the American Statistical Association* 81: 945–70.

(October 26, 1993). *Personal communication*. Princeton, NJ.

Hopkins, Eric. (1989). *Birmingham: The First Manufacturing Town in the World, 1760–1840*. London: Weidenfeld and Nicolson. http://www.theatlantic.com/education/archive/2013/12/the-case-for-sat-words/282253/ (accessed August 27, 2015).

Hume, D. (1740). A Treatise on Human Nature.

Huygens, C. (1669). In Huygens, C. (1895). *Oeuvres complétes, Tome Sixiéme Correspondance (pp.* 515–18, 526–39). The Hague, The Netherlands: Martinus Nijhoff.

Kahneman, D. (2012). *Thinking Fast, Thinking Slow*. New York: Farrar, Straus and Giroux.

Kalager, M., Zelen, M., Langmark, F., and Adami, H. (2010). Effect of screening mammography on breast-cancer mortality in Norway. *New England Journal of Medicine* 363: 1203–10.

Kant, I. (1960). *Religion within the Limits of Reason Alone (*pp. 83–4*)*. 2nd ed., trans. T. M. Green and H. H. Hudon. New York: Harper Torchbook.

Keranen, K. M., Savage, H. M., Abers, G. A., and Cochran, E. S. (June 2013). Potentially induced earthquakes in Oklahoma, USA: Links between wastewater injection and the 2011 $M_w$ 5.7 earthquake sequence. *Geology* 41: 699–702.

Keranen, K. M., Weingarten, M., Abers, G. A., Bekins, B. A., and Ge, S. (July 25, 2014). Sharp increase in central Oklahoma seismicity since 2008 induced by massive wastewater injection. *Science* 345(6195): 448–51. Published online July 3, 2014.

Kitahara, C. M., et al. (July 8, 2014). Association between class III obesity (BMI of 40–59 kg/m) and mortality: A pooled analysis of 20 prospective studies. *PLOS Medicine*. doi: 10.1371/journal.pmed.1001673.

Kolata, G. (July 8, 2013). What does birth cost? Hard to tell. *New York Times*.

Laplace, P. S. (1786). Sur les naissances, les mariages et les morts, á Paris, depuis1771 jusqui'en 1784 et dans tout l'entendue de la France, pendant les années 1781 et 1782. *Mémoires de l'Académie Royale des Sciences de Paris 1783*.

Little, R. J. A., and Rubin, D. B. (1987). *Statistical Analysis with Missing Data*. New York: Wiley.

Ludwig, D. S., and Friedman, M. I. (2014). Increasing adiposity: Consequence or cause of overeating? *Journal of the American Medical Association*. Published online: May 16, 2014. doi:10.1001/jama.2014.4133.

Luhrmann, T. M. (July 27, 2014). Where reason ends and faith begins. *New York Times*, News of the Week in Review, p. 9.

Macdonell, A. A. (January 1898). The origin and early history of chess. *Journal of the Royal Asiatic Society* 30(1): 117–41.

Mee, J., Clauser, B., and Harik, P. (April 2003). An examination of the impact of computer interruptions on the scores from computer administered examinations. Round table discussion presented at the annual meeting of the National Council of Educational Measurement, Chicago.

Meier, P. (1977). The biggest health experiment ever: The 1954 field trial of the Salk Poliomyelitis vaccine. In *Statistics: A Guide to the Study of the Biological and Health Sciences (pp.* 88–100). New York: Holden-Day.

Messick, S., and Jungeblut, A. (1981). Time and method in coaching for the SAT. *Psychological Bulletin* 89: 191–216.

Mislevy, R. J., Steinberg, L. S., and Almond, R. G. (2003). On the structure of educational assessments. *Measurement: Interdisciplinary Research and Perspectives* 1(1): 3–67.

Moore, A. (2010, August 27). Wyoming Department of Education Memorandum Number 2010–151: *Report on Effects of 2010 PAWS Administration Irregularities on Students Scores*. Retrieved on 30 August, 2013 from http://edu.wyoming.gov/PublicRelationsArchive/supt_memos/2010/2010_151.pdf.

Mosteller, F. (1995). The Tennessee study of class size in the early school grades. *The Future of Children* 5(2): 113–27.

Murphy, J. S. (December 11, 2013). The case for SAT Words. *The Atlantic*.

National Institutes of Health. (2014). Estimates of Funding for Various Research, Condition, and Disease Categories. http://report.nih.gov/categorical_spending.aspx (accessed September 29, 2014).

Neyman, J. (1923). On the application of probability theory to agricultural experiments. Translation of excerpts by D. Dabrowska and T. Speed. *Statistical Science* 5 (1990): 462–72.

Nightingale, F. (1858). *Notes on Matters Affecting the Health, Efficiency and Hospital Administration of the British Army*. London: Harrison and Sons.

Pacioli, Luca (1494). *Summa de Arithmetica*. Venice, folio 181, p. 44.

Pfeffermann, D., and Tiller, R. (2006). Small-area estimation with state-space models subject to benchmark constraints. *Journal of the American Statistical Association* 101: 1387–97.

Playfair, W. (1821). *A Letter on Our Agricultural Distresses, Their Causes and Remedies*. London: W. Sams.

(1801/2005). *The Statistical Breviary; Shewing on a Principle Entirely New, the Resources of Every State and Kingdom in Europe; Illustrated with Stained*

*Copper-Plate Charts, Representing the Physical Powers of Each Distinct Nation with Ease and Perspicuity*. Edited and introduced by Howard Wainer and Ian Spence. New York: Cambridge University Press.

(1786/1801). *The Commercial and Political Atlas, Representing, by Means of Stained Copper-Plate Charts, the Progress of the Commerce, Revenues, Expenditure, and Debts of England, during the Whole of the Eighteenth Century*. Facsimile reprint edited and annotated by Howard Wainer and Ian Spence. New York: Cambridge University Press, 2006.

Quinn, P. D., and Duckworth, A. L. (May 2007). Happiness and academic achievement: Evidence for reciprocal causality. Poster session presented at the meeting of the Association for Psychological Science, Washington, DC.

Reckase, M. D. (2006). Rejoinder: Evaluating standard setting methods using error models proposed by Schulz. *Educational Measurement: Issues and Practice* 25(3): 14–17.

Robbins, A. (2006). *The Overachievers: The Secret Lives of Driven Kids*. New York: Hyperion.

Robinson, A. H. (1982). *Early Thematic Mapping in the History of Cartography*. Chicago: University of Chicago Press.

Rosen, G. (February 18, 2007). Narrowing the religion gap. *New York Times Sunday Magazine*, p. 11.

Rosenbaum, P. (2009). *Design of Observational Studies*. New York: Springer.

(2002). *Observational Studies*. New York: Springer.

Rosenthal, J. A., Lu, X., and Cram, P. (2013). Availability of consumer prices from US hospitals. *JAMA Internal Medicine* 173(6): 427–32.

Rubin, D. B. (2006). Causal inference through potential outcomes and principal stratification: Application to studies with "censoring" due to death. *Statistical Science* 21(3): 299–309.

(2005). Causal inference using potential outcomes: Design, modeling, decisions. 2004 Fisher Lecture. *Journal of the American Statistical Association* 100: 322–31.

(1978). Bayesian inference for causal effects: The role of randomization. *The Annals of Statistics* 7: 34–58.

(1975). Bayesian inference for causality: The importance of randomization. In Social Statistics Section, *Proceedings of the American Statistical Association*: 233–9.

(1974). Estimating causal effects of treatments in randomized and non-randomized studies. *Journal of Educational Psychology* 66: 688–701.

Scherr, G. H. (1983). Irreproducible science: Editor's introduction. In *The Best of the Journal of Irreproducible Results*. New York: Workman Publishing.

Sinharay, S. (2014). Analysis of added value of subscores with respect to classification. *Journal of Educational Measurement* 51(2): 212–22.

(2010). How often do subscores have added value? Results from operational and simulated data. *Journal of Educational Measurement* 47(2): 150–74.

Sinharay, S., Haberman, S. J., and Wainer, H. (2011). Do adjusted subscores lack validity? Don't blame the messenger. *Educational and Psychological Measurement* 7(5): 789–97.

Sinharay, S., Wan, P., Whitaker, M., Kim, D-I., Zhang, L., and Choi, S. (2014). Study of the overall impact of interruptions during online testing on the test scores. Unpublished manuscript.

Slavin, Steve (1989). *All the Math You'll Ever Need* (pp. 153–4). New York: John Wiley and Sons.

Solochek, J. (2011, May 17). Problems, problems everywhere with Pearson's testing system. *Tampa Bay Times*. Retrieved on 30 August, 2013 from http://www.tampabay.com/blogs/gradebook/content/problems-problems-everywhere-pearsons-testing-system/2067044.

Strayer, D. L., Drews, F. A., and Crouch, D. J. (2003). Fatal distraction? A comparison of the cell-phone driver and the drunk driver. In D. V. McGehee, J. D. Lee, and M. Rizzo (Eds.), *Driving Assessment 2003: International Symposium on Human Factors in Driver Assessment, Training, and Vehicle Design* (pp. 25–30). Public Policy Center, University of Iowa.

Thacker, A. (2013). Oklahoma interruption investigation. Presented to the Oklahoma State Board of Education.

Thissen, D., and Wainer, H. (2001). *Test Scoring*. Hillsdale, NJ: Lawrence Erlbaum Associates.

Tufte, E. R. (2006). *Beautiful Evidence*. Cheshire, CT: Graphics Press.

(November 15, 2000). Lecture on information display given as part of the Yale Graduate School's Tercentennial lecture series "In the Company of Scholars" at Levinson Auditorium of the Yale University Law School.

(1996). *Visual Explanations*. Cheshire, CT: Graphics Press.

(1990). *Envisioning Information*. Cheshire, CT: Graphics Press.

(1983/2000). *The Visual Display of Quantitative Information*. Cheshire, CT: Graphics Press.

Twain, M. (1883). *Life on the Mississippi*. Montreal: Dawson Brothers.

Verkuyten, M., and Thijs, J. (2002). School satisfaction of elementary school children: The role of performance, peer relations, ethnicity, and gender. *Social Indicators Research* 59(2): 203–28.

Wainer, H. (2012). Moral statistics and the thematic maps of Joseph Fletcher. *Chance* 25(1): 43–7.

(2011a). *Uneducated Guesses Using Evidence to Uncover Misguided Education Policies*. Princeton, NJ: Princeton University Press.

(2011b). Value-added models to evaluate teachers: A cry for help. *Chance* 24(1): 11–13.

(2009). *Picturing the Uncertain World: How to Understand, Communicate and Control Uncertainty through Graphical Display*. Princeton, NJ: Princeton University Press.

(2007). Galton's normal is too platykurtic. *Chance* 20(2): 57–8.

(2005). *Graphic Discovery: A Trout in the Milk and Other Visual Adventures.* Princeton, NJ: Princeton University Press.

(2002). Clear thinking made visible: Redesigning score reports for students. *Chance* 15(1): 56–8.

(2000a). Testing the disabled: Using statistics to navigate between the Scylla of standards and the Charybdis of court decisions. *Chance* 13(2): 42–4.

(2000b). *Visual Revelations: Graphical Tales of Fate and Deception from Napoleon Bonaparte to Ross Perot.* 2nd ed. Hillsdale, NJ: Lawrence Erlbaum Associates.

(1984). How to display data badly. *The American Statistician* 38: 137–47.

(1983). Multivariate displays. In M. H. Rizvi, J. Rustagi, and D. Siegmund (Eds.), *Recent Advances in Statistics (*pp. 469–508). New York: Academic Press.

Wainer, H., and Rubin, D. B. (2015). Causal inference and death. *Chance* 28(2): 54–62.

Wainer, H., Lukele, R., and Thissen, D. (1994). On the relative value of multiple-choice, constructed response, and examinee-selected items on two achievement tests. *Journal of Educational Measurement* 31: 234–50.

Wainer, H., Sheehan, K., and Wang, X. (2000). Some paths toward making Praxis scores more useful. *Journal of Educational Measurement* 37: 113–40.

Wainer, H., Bridgeman, B., Najarian, M., and Trapani, C. (2004). How much does extra time on the SAT help? *Chance* 17(2): 19–24.

Walker, C. O., Winn, T. D., and Lutjens, R. M. (2008). Examining relationships between academic and social achievement goals and routes to happiness. *Education Research International* (2012), Article ID 643438, 7 pages. http://dx.doi.org/10.1155/2012/643438 (accessed August 27, 2015).

Waterman, A. S. (1993). Two conceptions of happiness: Contrasts of personal expressiveness (eudaimonia) and hedonic enjoyment. *Journal of Personality and Social Psychology* 64(4): 678–91.

Wilkinson, L. (2005). *The Grammar of Graphics.* 2nd ed. New York: Springer-Verlag.

Winchester, S. (2009). *The Map That Changed the World.* New York: Harper Perennial.

Wyld, James (1815) in Jarcho, S. (1973). Some early demographic maps. *Bulletin of the New York Academy of Medicine* 49: 837–44.

Yowell, T., and Devine, J. (May 2014). Evaluating current and alternative methods to produce 2010 county population estimates. U.S. Census Bureau Working Paper No. 100.

Zahra S., Khak, A. A., and Alam, S. (2013). Correlation between the five-factor model of personality-happiness and the academic achievement of physical education students. *European Journal of Experimental Biology* 3(6): 422–6.

Zieky, M. J., Perie, M., and Livingston, S. (2008). Cutscores: A manual for setting standards of performance on educational and occupational tests. http://www.amazon.com/Cutscores-Standards-Performance-Educational-Occupational/dp/1438250304/ (accessed August 27, 2015).

# Sources

Chapter 1
How the Rule of 72 Can Provide Guidance to Advance Your Wealth, Your Career, and Your Gas Mileage. *Chance* 26(3), 47–8, 2013.

Chapter 2
Piano Virtuosos and the Four-Minute Mile. *Significance* 9(1), 28–9, 2012.

Chapter 3
Happiness and Causal Inference. *Chance* 27(4), 61–4, 2014.

Chapter 4
Causal Inference and Death (with D. B. Rubin). *Chance* 28(2), 58–64, 2015.

Chapter 5
Using Experiments to Answer Four Vexing Questions. *Chance* 29(1), forthcoming, 2016.

Chapter 7
Life Follows Art: Gaming the Missing Data Algorithm. *Chance* 27(2), 56–7, 2014.

Chapter 8
On the Crucial Role of Empathy in the Design of Communications: Testing as an Example. *Chance* 27(1), 45–50, 2014.

Chapter 9
Improving Data Displays: Our's and The Media's. *Chance* 20(3), 8–16, 2007.

Improving Data Displays: Our's and The Media's. Chapter 11 in Wainer, H. *Picturing the Uncertain World*. Princeton, NJ: Princeton University Press, 92–105, 2009.

Chapter 10
Inside Out Plots (with J. O. Ramsay). *Chance* 23(3), 57–62, 2010.

Chapter 11
A Century and a Half of Moral Statistics in the United Kingdom: Variations on Joseph Fletcher's Thematic Maps (with R. Cook). *Significance* 9(3), 31–6, 2012.

Moral Statistics and the Thematic Maps of Joseph Fletcher. *Chance* 25(1), 43–7, 2012.

Chapter 12
Waiting for Achilles. *Chance* 25(4), 50–1, 2012.

Chapter 13
How Much Is Tenure Worth? *Chance* 24(3), 54–7, 2011.

Chapter 14
Cheating: Some Ways to Detect It Badly. *Chance* 25(3), 54–7, 2012.

Cheating: Some Ways to Detect It Badly. Chapter 3 in Kingston, N., and Clark, A. K. *Test Fraud: Statistical Detection and Methodology*. New York: Routledge, 8–20, 2014.

Chapter 15
When Nothing Is Not Zero: A True Saga of Missing Data, Adequate Yearly Progress, and a Memphis Charter School. *Chance* 25(2), 49–51, 2012.

Chapter 16
Musing about Changes in the SAT: Is the College Board Getting Rid of the Bulldog? *Chance* 27(3), 59–63, 2014.

Chapter 17
For Want of a Nail: Why Worthless Subscores May Be Seriously Impeding the Progress of Western Civilization (with R. Feinberg). *Significance* 12(1), 16–21, 2015.

Conclusion
~~Don't~~ Try This at Home. *Chance* 29(2), forthcoming, 2016.

# Index